HUMAN ANCESTORS

Readings from
**SCIENTIFIC
AMERICAN**

HUMAN ANCESTORS

With an Introduction by
Glynn Isaac
University of California, Berkeley

Richard E. F. Leakey
National Museums of Kenya

W. H. Freeman and Company
San Francisco

All of the SCIENTIFIC AMERICAN articles in *Human Ancestors* are available as separate Offprints. For a complete list of articles now available as Offprints, write to W. H. Freeman and Company, 660 Market Street, San Francisco, California, 94104.

Library of Congress Cataloging in Publication Data

Main entry under title:

Human ancestors.

   Bibliography: p.
   Includes index.
   1. Human evolution—Addresses, essays, lectures.
2. Paleolithic period–Addresses, essays, lectures.
I. Isaac, Glynn L., 1937–   II. Leakey, Richard E.
III. Scientific American.
American.
GN281.H847   573.2   79–4486
ISBN 0–7167–1100–1
ISBN 0–7167–1101–X pbk.

Printed in the United States of America

9 8 7 6 5 4 3 2

PREFACE

What kind of beings are we humans? An age-old question to which, until comparatively recently, different answers were offered by different peoples all over the world. However, all of these answers possessed strong components of myth and mysticism. Most cultures on all continents wove poetic stories of creation into the fabric of their understanding of the nature of human nature. For many cultures the origin myths did far more than cope with curiosity about beginnings; they served to establish human identity and to inculcate a sense of destiny. Very often, creation myths treated the acquisition of the endowments that distinguish humans from animals—endowments such as speech, control of fire, or the knowledge of good and evil. Then one hundred and twenty years ago Charles Darwin's *Origin of Species* articulated an alternative to the myths and allegories that had hitherto been all but universal. This opened a new epoch in human thinking about humanity and established a new realm of scientific endeavor.

Darwin was not the first to consider the possibility that life, including human life, had originated through a prolonged process of gradual change involving natural rather than supernatural mechanisms. The idea is contained in some myths and classical writings, and various eighteenth and early nineteenth century naturalists, such as Linnaeus and Lamarck, were groping for it. Darwin was simply the first to articulate the theory with sufficient clarity to challenge existing beliefs. The new theory transformed people's thoughts about themselves in two ways. First, it implies that humans are part of a commonwealth of nature rather than being a special creation with dominion over nature. Second, the theory takes the story of human origins out of the realm of religious dogma and makes it a matter for scientific exploration and inquiry. The century of geological and biological research that followed established that all who accept the rules of scientific evidence must incorporate evolutionary derivation into their understanding of themselves.

This collection of articles from SCIENTIFIC AMERICAN describes various aspects of the search for evidence concerning human origins and development. Descriptions of 10- to 30-million-year-old fossils, which were definitely not human, provide clues to the nature of the ancestors that we share with our closest living relatives, the apes. Proto-human beings, like us in body form but lacking our mental capabilities, established patterns of life that have passed without leaving living counterparts.

The collection deals both with the paleontological record of successive changes in anatomic form and with the archaeological record of early stages in the development of technology, economy, and cultural elaboration. The articles suggest the ways in which environmental opportunities, behavior, body form, and brain function have all interacted in the process of change.

The collection shows how the growth of knowledge has accelerated over the past two decades and shows the excitement of searches and discoveries, but it also demonstrates how very incomplete our understanding yet remains.

The theory of evolution profoundly changes the specifics of our beliefs about our origins, but it does not allow us to escape the question "what kind of creatures are human beings?" What traits have we inherited from our past? Are we an inherently violent species, as some have argued, or do we have in our natures deep-rooted propensities for cooperative endeavor? We will need more evidence to understand the complex components of the processes that have formed us, but it is already clear that, like the origin myths they replace, paleontological histories of our species are not morally neutral. Through their concern with whence we came, the articles in this collection encourage the reader to consider what we are and what we might become.

Just as modern astronomy and astrophysics are producing a sense of the universe that is more awesome than the cosmos of the ancients, so the scientific exploration of the roots of humanity serves to underscore the extraordinary scope and intricacy of the qualities of human nature.

January 1979

Glynn Isaac

Richard E. F. Leakey

CONTENTS

Note on cross-references to SCIENTIFIC AMERICAN *articles:* Articles included in this book are referred to by title and page number; articles not included in this book but available as Offprints are referred to by title and offprint number; articles not included in this book and not available as Offprints are referred to by title and date of publication.

HUMAN ANCESTORS

INTRODUCTION

Just as the voyage of Columbus opened up a new continent for European exploration, so the insight of Charles Darwin made science aware of an uncharted realm. Darwin showed that the living organisms of the modern world were each the end product of a long, long process of change. Science was given the challenge of exploring the history of life. From the start, the charting of the source and the course of the evolutionary stream that led to the human species was an aspect that held special interest.

The recovery of the record of human descent as we know it 120 years later has involved a series of dramatic discoveries. The first major pre-*Homo sapiens* discovery occurred in 1891 when Eugène Dubois discovered Pithecanthropus fossils in Java, a story told in W. W. Howells' essay on *Homo erectus*. Two major discoveries followed during the 1920s. In quick succession the Pekin man fossils were found in China, and the first *Australopithecus* fossil was found, recognized, and named in Africa. This last find involved a less human form, and at that time many scientists refused to recognize that the relatively small-brained, two-legged *Australopithecus* was a close relative of the human species. Raymond Dart's view that *Australopithecus africanus* is literally an ancestral species is still debated, but very few scientists now doubt that human ancestors passed through a phase of being small-brained, two-legged, proto-human beings. In the 1930s, other discoveries were made that were not at the time appreciated as dramatic. These included finds in northern India of fossil teeth that were appreciably older than the *Australopithecus* fossils and that have a markedly human-like aspect. As Elwyn Simons explains in the second article, their discoverer, G. E. Lewis called them *Ramapithecus* and put them in the human family, Hominidae. Also in the 1930s, Robert Broom began to follow up Dart's discovery by searching for fossil-bearing caves in the Transvaal region of southern Africa. He achieved spectacular success, and his sites—Sterkfontein, Swartkrans, and Kromdraai—have yielded hundreds of specimens of several varieties of early human-like beings. Broom's discoveries, coupled with careful reports by the eminent anatomist, the late W. E. Le Gros Clark, served to vindicate many of Dart's claims and to focus world attention on Africa as the potential cradle of human evolution.

Meanwhile from the entire Old World, a collection of accidentally discovered fossil bits and pieces was accumulating, most of them isolated specimens. These finds included "Rhodesian Man" from Zambia; the Swanscombe skull, England; the Mauer jaw and Steinheim skull, Germany; and the Solo skulls, Indonesia. These and many other fossils provide valuable evidence, but they inform us mainly about the later chapters of the story, that is, the last few hundred thousand years. Sources of evidence regarding earlier stages were confined to the Transvaal, Java, and Pekin, and, much older, India. The early years of the twentieth century had also brought to light the Piltdown skull, which, after misleading many scientists, was proved to be a fake.

During the hundred years after Darwin, the growth of knowledge of human evolution was limited by a number of factors. First, the discovery of sources of evidence depended mainly on accidents, such as the finding of fossil human teeth in a Chinese drug store or the recovery by a worker of a skull from a lime quarry in South Africa. Second, not enough was known to make it feasible to choose places for controlled and productive search. Third, only a very small number of scientists in the whole world had human evolution as their major topic of interest. Then, about twenty years ago this field of science underwent a series of transformations. There began to be enough knowledge to choose places to look rather than waiting for accidental finds, and in many areas systematic searches had been and still are turning up numerous specimens and rich coherent bodies of contextual evidence. Involvement of diverse scientists in the field has increased more than tenfold.

The turning point on the growth curve of paleoanthropology was associated with a major discovery: that of "Zinjanthropus," made at Olduvai Gorge in 1959 by Mary Leakey. This find was itself the culmination of years of searching by Louis and Mary Leakey, who had held to a faith that the East African Rift was a very important potential repository of evidence. Louis and Mary Leakey's systematic excavation and searches at Olduvai continued to deliver such spectacular finds that they helped to stimulate a whole new research movement.

This collection of articles documents aspects of progress during these exciting last two decades. The collection must be viewed as a progress report: The earlier articles from the 1960s can already be seen as first approximations, which require revisions in some ways that will be pointed out further on in this introduction. No doubt in five or ten years the more recent articles will be subject to similar revision, but the research of these two decades has lifted the study of human origins to the threshold of maturity.

The article chosen to open the collection is "Tools and Human Evolution" by Sherwood Washburn. Published in 1960, just at the turning point discussed above, it provided at the outset of the new era of research a clear-minded discussion of those fundamental, distinctive human attributes that have arisen during the evolutionary development of our species, namely:

Locomotion on the two hind limbs (bipedalism) with the fore limbs free for nonlocomotor functions

Dependence for adaptation and survival on skillfully made tools and equipment

Enlargement and reorganization of the brain relative to the brains of other higher primates

Development of speech and language

Development of social patterns involving cultural controls on aggression and on sexual actions and involving the division of labor

The reader must take into account the need for various revisions that have been occasioned by subsequent research, particularly the age of the earlier fossils discussed in the article, which are now reliably estimated as three to four times older than the half to one million years shown in the charts. Equally, more recent evidence strongly suggests that even the earliest hominids discussed were fully bipedal. However, determining the sequence of these innovations and elucidating their adaptive interrelationships remain the central challenges faced by paleoanthropology. Readers will find these themes recurring throughout the collection.

Two aspects of the modern study of human evolution can usefully be recognized. First, there is the need to determine the *narrative* of change, and second,

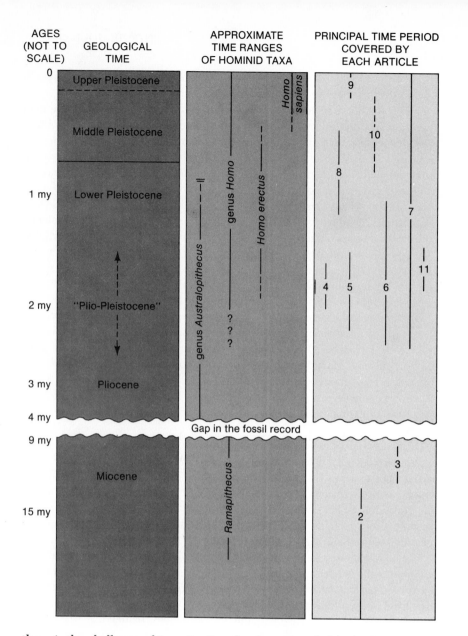

AGES (NOT TO SCALE)	GEOLOGICAL TIME	APPROXIMATE TIME RANGES OF HOMINID TAXA	PRINCIPAL TIME PERIOD COVERED BY EACH ARTICLE

there is the challenge of investigating the *dynamics* of change. By narrative we mean the sequence of changes in anatomical form, in subsistence, technological practices, and culture, and the sequence of changes in ecology and geography. None of these can ever be fully reconstructed from the fossil record, but valuable information can be gathered. By dynamics we mean the interplay of factors that caused evolutionary change to occur, which should be explicable in terms of selection, that is, differential gene survival. In practice, however, studying dynamics involves attempting to understand the main components of complex adaptive systems involving energy (food), reproduction, social organization, and environmental relationships.

Clearly, detailed studies of dynamics are difficult until the narrative is known in outline, so readers will find that most of the articles are primarily concerned with establishing the sequence of change. With the exception of the first and last, which do concern themselves more specifically with dynamics, the articles have been arranged in chronological sequence of topics covered. Elwyn Simons' "The Early Relatives of Man" is an exposition of paleontological knowledge about the primate stocks in the time range 60–10 million years ago. From these stocks the human lineage sprang. The papers then treat successively Miocene, Plio-Pleistocene, Middle Pleistocene, and later Pleistocene evidence. (See the chart on this page.)

Readers will find that at a certain point in the chronological sequence there is a change in content coverage. Articles 2–8 deal primarily with the evidence gained from the fossils, whereas articles 9–11 deal with archaeological evidence for economy, technology, and culture. This change in subject matter fairly reflects the fact that over the last span of human evolution the most profound changes were in the realm of behavior, whereas anatomical changes in the last million years have been slight.

A narrative can only be intelligible if the events of the story can be ordered in time, and for this reason one of the most important refinements of the past twenty years has been the development of a chronology based on dates achieved through measurements of amounts of radioactive decay. For the time period that concerns us here, dates are chiefly established by the potassium-argon and the fission-track methods. The application of these methods in areas where volcanic rocks are interbedded with sedimentary successions appears to have provided a generally consistent time scale, although in particular instances there are still uncertainties and debates. Alan Walker and Richard Leakey (article 6) present information on one such case. Considerable chronological uncertainty exists over some important sites: for instance, the Transvaal *Australopithecus* sites and the Indonesian (Java) *Homo erectus* locations of discovery. The chart (opposite page) indicates current estimates of the dates for boundaries between geological time divisions. Readers will find that some of the articles written before 1970 (1, 2, and 8) have rather different chronologies in their charts; this discrepancy should be viewed from a historical perspective. The chart also shows the approximate time covered in the subject matter of each article and gives an estimate of the time range covered by the main taxonomic entities referred to in the articles.

Although the articles are arranged in chronological order of the subject matter, there is another useful way of looking at the record, namely, to start with the present and to look back into the past, seeing in what segments of time familiar aspects of the modern situation first become evident in the record:

		Years before present
1.	Farming and food production	10–12,000
2.	Mechanical devices like bows and spear throwers	30,000
3.	Representational art	30,000
4.	Hafted tools appear	50–100,000
5.	Humans occupy the cold, temperate zone	300–500,000
6.	Definite signs of controlled fire	300–500,000
7.	Last definite signs of multiple species of coexistent hominids	1 million
8.	Oldest definite "camp sites"	2 m
9.	Oldest known definite stone artifacts	2–2.5 m
10.	Oldest fossil skull with a brain case much larger than that of any ape	2–2.5 m
11.	Oldest known definite indications of a fully bipedal mode of locomotion	3–4 m

At this point there is a virtual gap in the fossil record between about 4 and 9 million years ago, and only a few uninformative scraps have been found.

12.	Oldest known traces of hominoids with teeth and jaws like those of later hominids (at least three species: *Ramapithecus*, *Sivapithecus*, and *Gigantopithecus*)	9–14 m

If we trace the record back beyond this point, we reach a time range where no fossils seem to show more specific resemblance to humans than to apes.

This list can be read from the top down, in which case each item should be subtracted from our familiar sense of the human condition, or it can be read from the bottom up, in which case it is a cumulative list of innovations.

If we confine ourselves to the anatomical innovations with which the articles mostly deal and take the timetable of first evidence at its face value, the following sequence could be argued:

(a) A shift in tooth and jaw form, perhaps indicative of changes in habitat and diet, is evident in widespread hominoid fossils from between 10 and 14 million years ago. This is treated in Elwyn Simons' article on *Ramapithecus*. It is a matter of controversy as to whether any of the earliest fossils with human-like teeth are actually human ancestors, but, whatever the outcome of this debate, these dental resemblances are the first hominid-like features (as opposed to generalized hominoid features) to appear in the fossil record.

(b) By 4 million years ago, some hominoids had adopted fully bipedal locomotion, and all of these can definitely be classified as hominids. The articles by John Napier describe aspects of the anatomical and functional implications of this transition. Napier used what was, at the time, the oldest known dated fossil materials as his principal fossil evidence, the material from Bed I at Olduvai (1.8 million years old). Since these articles were written, still earlier fossil evidence for bipedalism as far back as 3 million years has been found by Don Johanson and Maurice Taieb at the Afar in Ethiopia.

(c) The next clear trends of anatomical change were partly concurrent. They involved (1) absolute and relative increase in brain size and (2) reduction in tooth and jaw size. These trends appear to have characterized one lineage more than others, presumably that leading to modern humans. The taxonomic complexities of documenting early stages of these trends is treated in Walker and Leakey. Ralph Holloway provides a statement of evidence regarding the brain volumes of hominid fossils of different ages and a more controversial claim that although many early fossil hominids had smaller brains than modern humans, the brain size differences are largely explicable in terms of the smaller body-size differences.

The record suggests that the onset of the strong trend to brain enlargement and the trend to tooth reduction was concomitant with stone tool-making. Presumably the making of tools from materials that were simpler to shape extends even further back in time. At much the same era, the oldest circumstantial evidence for meat-eating, for carrying, for the use of home bases, and for food-sharing appears in the record. In article 11, Glynn Isaac argues that a set of behaviors, including those mentioned above, proved adaptive and imposed strong selection pressures on brain function and the ability to communicate through language-like means. Through natural selection this behavioral complex, which would also involve social cooperation and increasingly complex culture, may well have entered into a positive feedback relationship with the evolution of the neural mechanisms of the brain. Much remains to be done to test hypotheses of this kind against specific evidence.

Stone technology is the aspect of human cultural behavior for which we have the longest record, and it consequently provides an important potential measure of those developing technical and cultural capabilities that are only distantly intimated by the fossils. However, for a very long time, stone tools were studied with a cultural-historic frame of reference that is more appropriate for Greek vases and other later artifacts. The article by Sally and Lewis

Binford helped to revolutionize studies of early artifacts by developing the argument that the variation between sets of stone tools at different sites is more likely to reflect the pursuit of different activities with different tool needs, than to reflect occupancy by members of different tribes. They use material from the Mousterian of Europe and the Near East to illustrate the argument. This material is about 40,000–100,000 years old. Archaeologists working on stone tool assemblages of all ages are still testing this important hypothesis, seeking to use variation as a key to the development of economic and land use patterns.

One of the most serious limitations in our understanding of the long-term record of stone-tool making has been almost complete ignorance of the usage to which the tools were put and of their role in adaptation. Lawrence Keeley's article reports on a very promising technique that may help overcome this limitation.

This anthology illustrates well that what was the terra incognita of human ancestry only 120 years ago has now been partially explored. However, it should also be understood that there are gaps, and many interpretations should be seen as tentative. For instance, from the time range 15–9 million years ago no definitely human form has yet been discovered; but very human-like teeth have been found that seem to belong to ground-dwelling creatures that were not necessarily familiar apes or humans. Several species of such creatures seem to have existed, but one, *Ramapithecus*, has particularly "human" dentition, and it is tempting to identify this taxon as an ancestor. This is an hypothesis, and some scientists have challenged the interpretation on the grounds that the close similarity of the biochemistry of humans and the African apes makes an evolutionary divergence date as remote as 15 million years very improbable. This debate can probably not be resolved until evidence is found from the gap. At present we have almost no specimens bearing on human ancestry from between 9 and 4 million years ago. But the hunt is on for fossil-bearing localities to fill this gap, and in coming years we can expect the excitement of such discoveries.

As Walker and Leakey make clear in article 6, in the period following the gap, the comparatively small and geographically restricted sets of fossil samples that we do have show an embarrassing diversity of forms. It is as though the basic protohuman anatomical and adaptive pattern had proved so successful that a number of species had arisen. But the fossils do not come with labels! The matter of how many species there were at any one time is a question of interpretation. Equally, scientists differ in their judgment as to how the forms from successive ages can best be linked as representatives of several evolving lineages. They debate which, if any, of these lineages is our own ancestral line. The evolutionary tree diagrams in all the articles must be understood as working hypotheses, not as established truth.

As we said at the outset of this introduction, this collection of articles can be seen as illustrating the progress of exploration during the past two decades. During this brief period, major fossil discoveries have been made at an ever-increasing rate, and the systematic investigation of their ecological and behavioral context has steadily intensified. The pursuit of knowledge about human ancestry is rapidly becoming a more and more truly international endeavor, and we can expect the tempo to quicken even further in the immediate future.

Tools and Human Evolution

by Sherwood L. Washburn
September 1960

*It is now clear that tools antedate man, and that their
use by prehuman primates gave rise to* Homo sapiens

A series of recent discoveries has linked prehuman primates of half a million years ago with stone tools. For some years investigators had been uncovering tools of the simplest kind from ancient deposits in Africa. At first they assumed that these tools constituted evidence of the existence of large-brained, fully bipedal men. Now the tools have been found in association with much more primitive creatures, the not-fully bipedal, small-brained near-men, or man-apes. Prior to these finds the prevailing view held that man evolved nearly to his present structural state and then discovered tools and the new ways of life that they made possible. Now it appears that man-apes—creatures able to run but not yet walk on two legs, and with brains no larger than those of apes now living—had already learned to make and to use tools. It follows that the structure of modern man must be the result of the change in the terms of natural selection that came with the tool-using way of life.

The earliest stone tools are chips or simple pebbles, usually from river

gravels. Many of them have not been shaped at all, and they can be identified as tools only because they appear in concentrations, along with a few worked pieces, in caves or other locations where no such stones naturally occur. The huge advantage that a stone tool gives to its user must be tried to be appreciated. Held in the hand, it can be used for pounding, digging or scraping. Flesh and bone can be cut with a flaked chip, and what would be a mild blow with the fist becomes lethal with a rock in the hand. Stone tools can be employed, moreover, to make tools of other materials. Naturally occurring sticks are nearly all rotten, too large, or of inconvenient shape; some tool for fabrication is essential for the efficient use of wood. The utility of a mere pebble seems so limited to the user of modern tools that it is not easy to comprehend the vast difference that separates the tool-user from the ape which relies on hands and teeth alone. Ground-living monkeys dig out roots for food, and if they could use a stone or a stick, they might easily double their food supply. It was the success of the simplest tools that started the whole trend of human evolution and led to the civilizations of today.

From the short-term point of view, human structure makes human behavior possible. From the evolutionary point of view, behavior and structure form an interacting complex, with each change in one affecting the other. Man began when populations of apes, about a mil-

lion years ago, started the bipedal, tool-using way of life that gave rise to the man-apes of the genus *Australopithecus*. Most of the obvious differences that distinguish man from ape came after the use of tools.

The primary evidence for the new view of human evolution is teeth, bones and tools. But our ancestors were not fossils; they were striving creatures, full of rage, dominance and the will to live. What evolved was the pattern of life of intelligent, exploratory, playful, vigorous primates; the evolving reality was a succession of social systems based upon the motor abilities, emotions and intelligence of their members. Selection produced new systems of child care, maturation and sex, just as it did alterations in the skull and the teeth. Tools, hunting, fire, complex social life, speech, the human way and the brain evolved together to produce ancient man of the genus *Homo* about half a million years ago. Then the brain evolved under the pressures of more complex social life until the species *Homo sapiens* appeared perhaps as recently as 50,000 years ago.

With the advent of *Homo sapiens* the tempo of technical-social evolution quickened. Some of the early types of tool had lasted for hundreds of thousands of years and were essentially the same throughout vast areas of the African and Eurasian land masses. Now the tool forms multiplied and became regionally diversified. Man invented the

STENCILED HANDS in the cave of Gargas in the Pyrenees date back to the Upper Paleolithic of perhaps 30,000 years ago. Aurignacian man made the images by placing hand against wall and spattering it with paint. Hands stenciled in black (*top*) are more distinct and apparently more recent than those done in other colors (*center*).

OLDUVAI GORGE in Tanganyika is the site where the skull of the largest known man-ape was discovered in 1959 by L. S. B. Leakey and his wife Mary. Stratigraphic evidence indicates that skull dates back to Lower Pleistocene, more than 500,000 years ago.

bow, boats, clothing; conquered the Arctic; invaded the New World; domesticated plants and animals; discovered metals, writing and civilization. Today, in the midst of the latest tool-making revolution, man has achieved the capacity to adapt his environment to his need and impulse, and his numbers have begun to crowd the planet.

The later events in the evolution of the human species are treated in other articles from the September, 1960 issue of SCIENTIFIC AMERICAN. This article is concerned with the beginnings of the process by which, as Theodosius Dobzhansky says in the concluding article of the issue, biological evolution has transcended itself. From the rapidly accumulating evidence it is now possible to speculate with some confidence on the manner in which the way of life made possible by tools changed the pressures of natural selection and so changed the structure of man.

Tools have been found, along with the bones of their makers, at Sterkfontein, Swartkrans and Kromdraai in South Africa and at Olduvai in Tanganyika. Many of the tools from Sterkfontein are merely unworked river pebbles, but someone had to carry them from the gravels some miles away and bring them to the deposit in which they are found. Nothing like them occurs naturally in the local limestone caves. Of course the association of the stone tools with man-ape bones in one or two localities does not prove that these animals made the tools. It has been argued that a more advanced form of man, already present, was the toolmaker. This argument has a familiar ring to students of human evolution. Peking man was thought too primitive to be a toolmaker; when the first manlike pelvis was found with man-ape bones, some argued that it must have fallen into the deposit because it was too human to be associated with the skull. In every case, however, the repeated discovery of the same unanticipated association has ultimately settled the controversy.

This is why the discovery by L. S. B. and Mary Leakey in the summer of 1959 is so important. In Olduvai Gorge in Tanganyika they came upon traces of an old living site, and found stone tools in clear association with the largest man-ape skull known. With the stone tools were a hammer stone and waste flakes from the manufacture of the tools. The deposit also contained the bones of rats, mice, frogs and some bones of juvenile pig and antelope, showing that even the largest and latest of the

SKULL IS EXAMINED *in situ* by Mary Leakey, who first noticed fragments of it protruding from the cliff face at left. Pebble tools were found at the same level as the skull.

SKULL IS EXCAVATED from surrounding rock with dental picks. Although skull was badly fragmented, almost all of it was recovered. Fragment visible here is part of upper jaw.

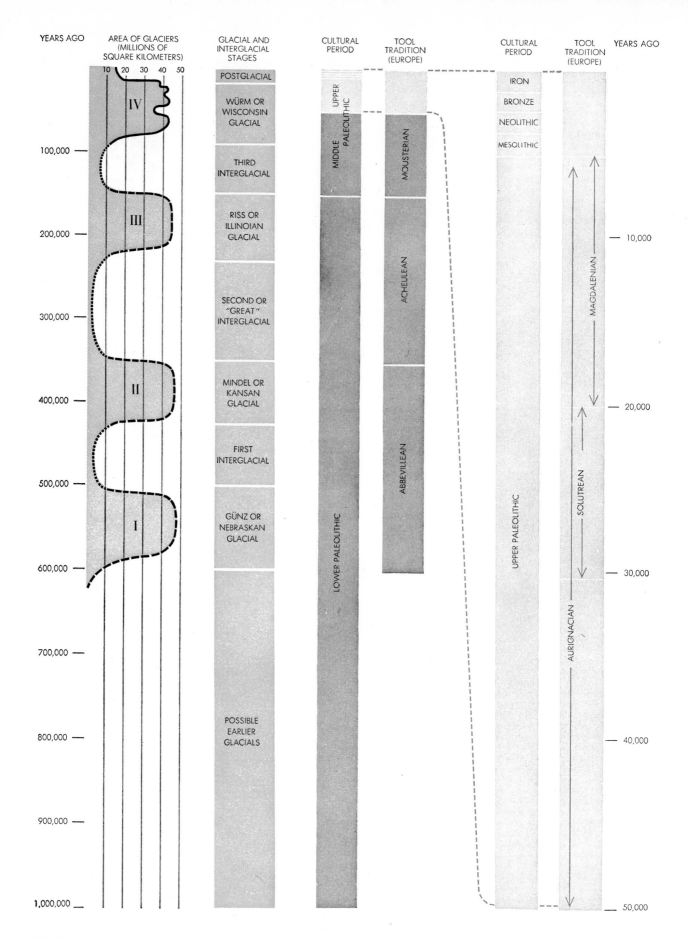

YEARS AGO	AREA OF GLACIERS (MILLIONS OF SQUARE KILOMETERS)	GLACIAL AND INTERGLACIAL STAGES	CULTURAL PERIOD	TOOL TRADITION (EUROPE)	CULTURAL PERIOD	TOOL TRADITION (EUROPE)	YEARS AGO

TIME-SCALE correlates cultural periods and tool traditions with the four great glaciations of the Pleistocene epoch. Glacial advances and retreats shown by solid black curve are accurately known; those shown by broken curve are less certain; those shown by dotted curve are uncertain. Light gray bars at far right show an expanded view of last 50,000 years on two darker bars at center. Scale was prepared with the assistance of William R. Farrand of the Lamont Geological Observatory of Columbia University.

man-apes could kill only the smallest animals and must have been largely vegetarian. The Leakeys' discovery confirms the association of the man-ape with pebble tools, and adds the evidence of manufacture to that of mere association. Moreover, the stratigraphic evidence at Olduvai now for the first time securely dates the man-apes, placing them in the lower Pleistocene, earlier than 500,000 years ago and earlier than the first skeletal and cultural evidence for the existence of the genus Homo [see illustration on next two pages]. Before the discovery at Olduvai these points had been in doubt.

The man-apes themselves are known from several skulls and a large number of teeth and jaws, but only fragments of the rest of the skeleton have been preserved. There were two kinds of man-ape, a small early one that may have weighed 50 or 60 pounds and a later and larger one that weighed at least twice as much. The differences in size and form between the two types are quite comparable to the differences between the contemporary pygmy chimpanzee and the common chimpanzee.

Pelvic remains from both forms of man-ape show that these animals were bipedal. From a comparison of the pelvis of ape, man-ape and man it can be seen that the upper part of the pelvis is much wider and shorter in man than in the ape, and that the pelvis of the man-ape corresponds closely, though not precisely, to that of modern man [see top illustration on page 17]. The long upper pelvis of the ape is characteristic of most mammals, and it is the highly specialized, short, wide bone in man that makes possible the human kind of bipedal locomotion. Although the man-ape pelvis is apelike in its lower part, it approaches that of man in just those features that distinguish man from all other animals. More work must be done before this combination of features is fully understood. My belief is that bipedal running, made possible by the changes in the upper pelvis, came before efficient bipedal walking, made possible by the changes in the lower pelvis. In the man-ape, therefore, the adaptation to bipedal locomotion is not yet complete. Here, then, is a phase of human evolution characterized by forms that are mostly bipedal, small-brained, plains-living, tool-making hunters of small animals.

The capacity for bipedal walking is primarily an adaptation for covering long distances. Even the arboreal chimpanzee can run faster than a man, and any monkey can easily outdistance him.

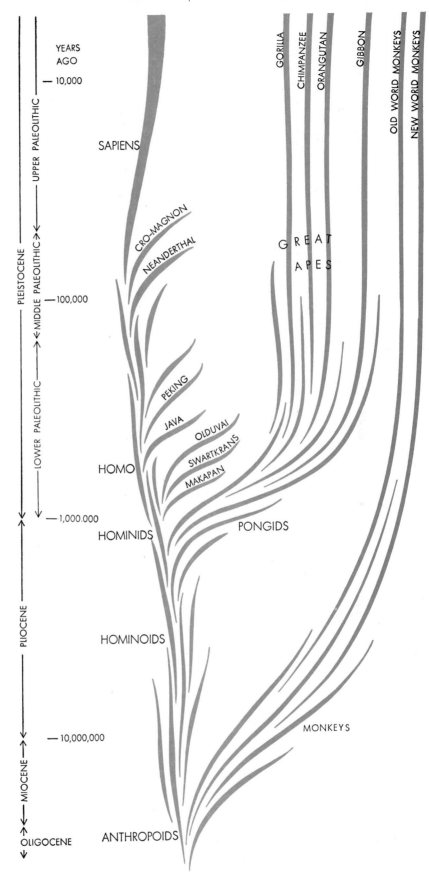

LINES OF DESCENT that lead to man and his closer living relatives are charted. The hominoid superfamily diverged from the anthropoid line in the Miocene period some 20 million years ago. From the hominoid line came the tool-using hominids at the beginning of the Pleistocene. The genus *Homo* appeared in the hominid line during the first interglacial (*see chart on opposite page*); the species *Homo sapiens*, around 50,000 years ago.

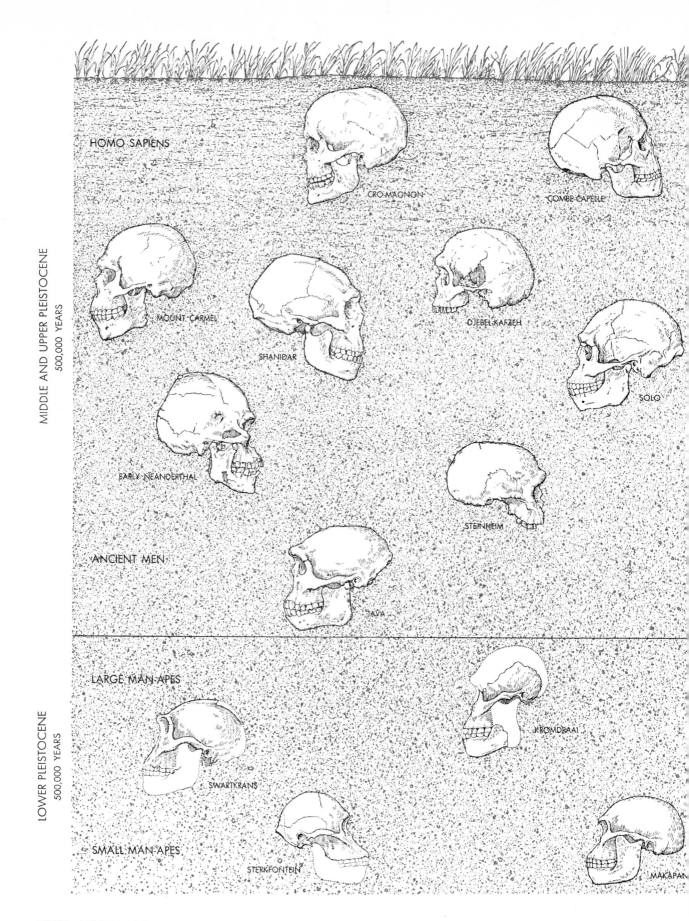

MIDDLE AND UPPER PLEISTOCENE
500,000 YEARS

LOWER PLEISTOCENE
500,000 YEARS

HOMO SAPIENS

CRO-MAGNON

COMBE CAPELLE

MOUNT CARMEL

SHANIDAR

DJEBEL KAFZEH

SOLO

EARLY NEANDERTHAL

STEINHEIM

ANCIENT MEN

JAVA

LARGE MAN-APES

KROMDRAAI

SWARTKRANS

SMALL MAN-APES

STERKFONTEIN

MAKAPAN

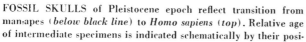

FOSSIL SKULLS of Pleistocene epoch reflect transition from man-apes (*below black line*) to *Homo sapiens* (*top*). Relative age of intermediate specimens is indicated schematically by their posi- tion on page. Java man (*middle left*) and Solo man (*upper center*) are members of the genus *Pithecanthropus*, and are related to Peking man (*middle right*). The Shanidar skull (*upper left*) be-

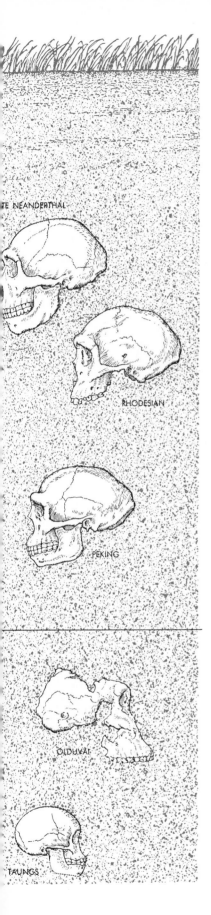

longs to the Neanderthal family, while Mount Carmel skull shows characteristics of Neanderthal and modern man.

A man, on the other hand, can walk for many miles, and this is essential for efficient hunting. According to skeletal evidence, fully developed walkers first appeared in the ancient men who inhabited the Old World from 500,000 years ago to the middle of the last glaciation. These men were competent hunters, as is shown by the bones of the large animals they killed. But they also used fire and made complicated tools according to clearly defined traditions. Along with the change in the structure of the pelvis, the brain had doubled in size since the time of the man-apes.

The fossil record thus substantiates the suggestion, first made by Charles Darwin, that tool use is both the cause and the effect of bipedal locomotion. Some very limited bipedalism left the hands sufficiently free from locomotor functions so that stones or sticks could be carried, played with and used. The advantage that these objects gave to their users led both to more bipedalism and to more efficient tool use. English lacks any neat expression for this sort of situation, forcing us to speak of cause and effect as if they were separated, whereas in natural selection cause and effect are interrelated. Selection is based on successful behavior, and in the man-apes the beginnings of the human way of life depended on both inherited locomotor capacity and on the learned skills of tool-using. The success of the new way of life based on the use of tools changed the selection pressures on many parts of the body, notably the teeth, hands and brain, as well as on the pelvis. But it must be remembered that selection was for the whole way of life.

In all the apes and monkeys the males have large canine teeth. The long upper canine cuts against the first lower premolar, and the lower canine passes in front of the upper canine. This is an efficient fighting mechanism, backed by very large jaw muscles. I have seen male baboons drive off cheetahs and dogs, and according to reliable reports male baboons have even put leopards to flight. The females have small canines, and they hurry away with the young under the very conditions in which the males turn to fight. All the evidence from living monkeys and apes suggests that the male's large canines are of the greatest importance to the survival of the group, and that they are particularly important in ground-living forms that may not be able to climb to safety in the trees. The small, early man-apes lived in open plains country, and yet none of them had large canine teeth. It would appear that the protection of the group must have shifted from teeth to tools early in the evolution of the man-apes, and long before the appearance of the forms that have been found in association with stone tools. The tools of Sterkfontein and Olduvai represent not the beginnings of tool use, but a choice of material and knowledge in manufacture which, as is shown by the small canines of the man-apes that deposited them there, derived from a long history of tool use.

Reduction in the canine teeth is not a simple matter, but involves changes in the muscles, face, jaws and other parts of the skull. Selection builds powerful neck muscles in animals that fight with their canines, and adapts the skull to the action of these muscles. Fighting is not a matter of teeth alone, but also of seizing, shaking and hurling an enemy's body with the jaws, head and neck. Reduction in the canines is therefore accompanied by a shortening in the jaws, reduction in the ridges of bone over the eyes and a decrease in the shelf of bone in the neck area [see illustration on page 18]. The reason that the skulls of the females and young of the apes look more like man-apes than those of adult males is that, along with small canines, they have smaller muscles and all the numerous structural features that go along with them. The skull of the man-ape is that of an ape that has lost the structure for effective fighting with its teeth. Moreover, the man-ape has transferred to its hands the functions of seizing and pulling, and this has been attended by reduction of its incisors. Small canines and incisors are biological symbols of a changed way of life; their primitive functions are replaced by hand and tool.

The history of the grinding teeth—the molars—is different from that of the seizing and fighting teeth. Large size in any anatomical structure must be maintained by positive selection; the selection pressure changed first on the canine teeth and, much later, on the molars. In the man-apes the molars were very large, larger than in either ape or man. They were heavily worn, possibly because food dug from the ground with the aid of tools was very abrasive. With the men of the Middle Pleistocene, molars of human size appear along with complicated tools, hunting and fire.

The disappearance of brow ridges and the refinement of the human face may involve still another factor. One of the essential conditions for the organi-

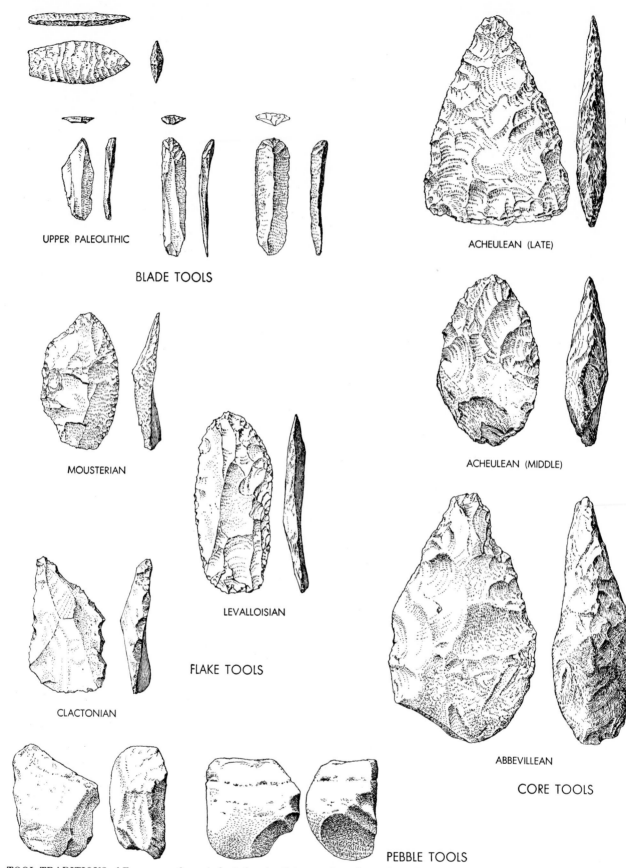

UPPER PALEOLITHIC

BLADE TOOLS

ACHEULEAN (LATE)

MOUSTERIAN

ACHEULEAN (MIDDLE)

LEVALLOISIAN

FLAKE TOOLS

CLACTONIAN

ABBEVILLEAN

CORE TOOLS

PEBBLE TOOLS

TOOL TRADITIONS of Europe are the main basis for classifying Paleolithic cultures. The earliest tools are shown at bottom of page; later ones, at top. The tools are shown from both the side and the edge, except for blade tools, which are shown in three views. Tools consisting of a piece of stone from which a few flakes have been chipped are called core tools (*right*). Other types of tool were made from flakes (*center and left*); blade tools were made from flakes with almost parallel sides. Tool traditions are named for site where tools of a given type were discovered; Acheulean tools, for example, are named for St. Acheul in France.

zation of men in co-operative societies was the suppression of rage and of the uncontrolled drive to first place in the hierarchy of dominance. Curt P. Richter of Johns Hopkins University has shown that domestic animals, chosen over the generations for willingness to adjust and for lack of rage, have relatively small adrenal glands. But the breeders who selected for this hormonal, physiological, temperamental type also picked, without realizing it, animals with small brow ridges and small faces. The skull structure of the wild rat bears the same relation to that of the tame rat as does the skull of Neanderthal man to that of *Homo sapiens*. The same is true for the cat, dog, pig, horse and cow; in each case the wild form has the larger face and muscular ridges. In the later stages of human evolution, it appears, the self-domestication of man has been exerting the same effects upon temperament, glands and skull that are seen in the domestic animals.

Of course from man-ape to man the brain-containing part of the skull has also increased greatly in size. This change is directly due to the increase in the size of the brain: as the brain grows, so grow the bones that cover it. Since there is this close correlation between brain size and bony brain-case, the brain size of the fossils can be estimated. On the scale of brain size the man-apes are scarcely distinguishable from the living apes, although their brains may have been larger with respect to body size. The brain seems to have evolved rapidly, doubling in size between man-ape and man. It then appears to have increased much more slowly; there is no substantial change in gross size during the last 100,000 years. One must remember, however, that size alone is a very crude indicator, and that brains of equal size may vary greatly in function. My belief is that although the brain of *Homo sapiens* is no larger than that of Neanderthal man, the indirect evidence strongly suggests that the first *Homo sapiens* was a much more intelligent creature.

The great increase in brain size is important because many functions of the brain seem to depend on the number of cells, and the number increases with volume. But certain parts of the brain have increased in size much more than others. As functional maps of the cortex of the brain show, the human sensory-motor cortex is not just an enlargement of that of an ape [*see illustrations on last three pages of this article*]. The areas for the hand, especially the thumb, in

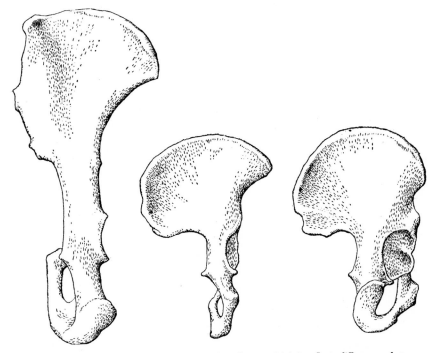

HIP BONES of ape (*left*), man-ape (*center*) and man (*right*) reflect differences between quadruped and biped. Upper part of human pelvis is wider and shorter than that of apes. Lower part of man-ape pelvis resembles that of ape; upper part resembles that of man.

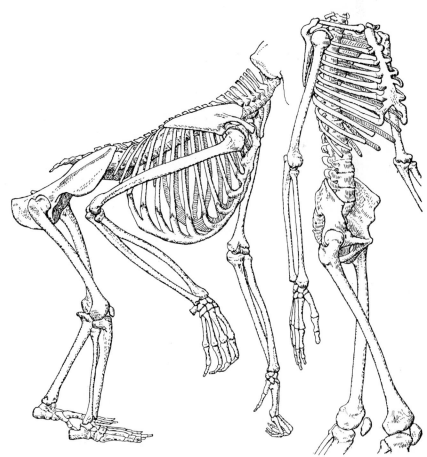

POSTURE of gorilla (*left*) and man (*right*) is related to size, shape and orientation of pelvis. Long, straight pelvis of ape provides support for quadrupedal locomotion; short, broad pelvis of man curves backward, carrying spine and torso in bipedal position.

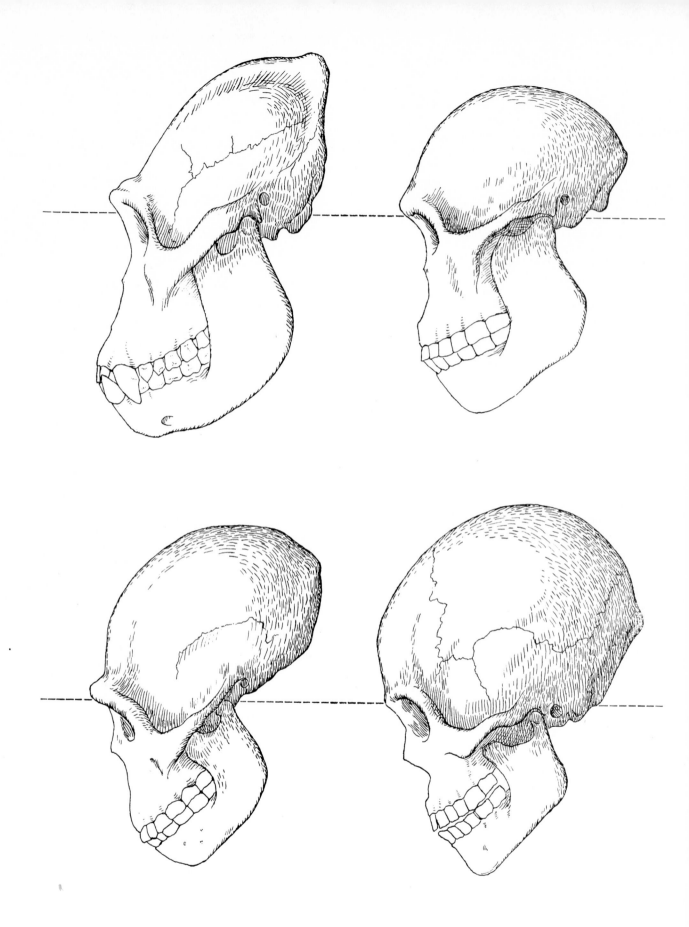

EVOLUTION OF SKULL from ape (*upper left*) to man-ape (*upper right*) to ancient man (*lower left*) to modern man (*lower right*) involves an increase in size of brain case (*part of skull above broken lines*) and a corresponding decrease in size of face (*part of skull below broken lines*). Apes also possess canine teeth that are much larger than those found in either man-apes or man.

for the hand, especially the thumb, in man are tremendously enlarged, and this is an integral part of the structural base that makes the skillful use of the hand possible. The selection pressures that favored a large thumb also favored a large cortical area to receive sensations from the thumb and to control its motor activity. Evolution favored the development of a sensitive, powerful, skillful thumb, and in all these ways —as well as in structure—a human thumb differs from that of an ape.

The same is true for other cortical areas. Much of the cortex in a monkey is still engaged in the motor and sensory functions. In man it is the areas adjacent to the primary centers that are most expanded. These areas are concerned with skills, memory, foresight and language; that is, with the mental faculties that make human social life possible. This is easiest to illustrate in the field of language. Many apes and monkeys can make a wide variety of sounds. These sounds do not, however, develop into language [see "The Origin of Speech," by Charles F. Hockett Offprint 603.] Some workers have devoted great efforts, with minimum results, to trying to teach chimpanzees to talk. The reason is that there is little in the brain to teach. A human child learns to speak with the greatest ease, but the storage of thousands of words takes a great deal of cortex. Even the simplest language must have given great advantage to those first men who had it. One is tempted to think that language may have appeared together with the fine tools, fire and complex hunting of the large-brained men of the Middle Pleistocene, but there is no direct proof of this.

The main point is that the kind of animal that can learn to adjust to complex, human, technical society is a very different creature from a tree-living ape, and the differences between the two are rooted in the evolutionary process. The reason that the human brain makes the human way of life possible is that it is the result of that way of life. Great masses of the tissue in the human brain are devoted to memory, planning, language and skills, because these are the abilities favored by the human way of life.

The emergence of man's large brain occasioned a profound change in the plan of human reproduction. The human mother-child relationship is unique among the primates as is the use of tools. In all the apes and monkeys the baby clings to the mother; to be able to do so,

MOTOR CORTEX OF MONKEY controls the movements of the body parts outlined by the superimposed drawing of the animal (*color*). Gray lines trace the surface features of the left half of the brain (*bottom*) and part of the right half (*top*). Colored drawing is distorted in proportion to amount of cortex associated with functions of various parts of the body. Smaller animal in right half of brain indicates location of secondary motor cortex.

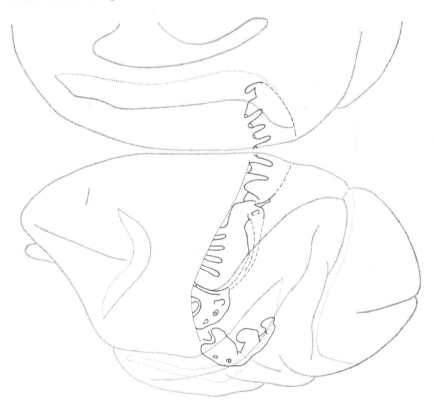

SENSORY CORTEX OF MONKEY is mapped in same way as motor cortex (*above*). As in motor cortex, a large area is associated with hands and feet. Smaller animal at bottom of left half of brain indicates location of secondary sensory cortex. Drawings are based on work of Clinton N. Woolsey and his colleagues at the University of Wisconsin Medical School.

the baby must be born with its central nervous system in an advanced state of development. But the brain of the fetus must be small enough so that birth may take place. In man adaptation to bipedal locomotion decreased the size of the bony birth-canal at the same time that the exigencies of tool use selected for larger brains. This obstetrical dilemma was solved by delivery of the fetus at a much earlier stage of development. But this was possible only because the mother, already bipedal and with hands free of locomotor necessities, could hold the helpless, immature infant. The small-brained man-ape probably developed in the uterus as much as the ape does; the human type of mother-child relation must have evolved by the time of the large-brained, fully bipedal humans of the Middle Pleistocene. Bipedalism, tool use and selection for large brains thus slowed human development and invoked far greater maternal responsibility. The slow-moving mother, carrying the baby, could not hunt, and the combination of the woman's obligation to care for slow-developing babies and the man's occupation of hunting imposed a fundamental pattern on the social organization of the human species.

As Marshall D. Sahlins suggests ["The Origin of Society," SCIENTIFIC AMERICAN Offprint 602], human society was heavily conditioned at the outset by still other significant aspects of man's sexual adaptation. In monkeys and apes year-round sexual activity supplies the social bond that unites the primate horde. But sex in these species is still subject to physiological—especially glandular—controls. In man these controls are gone, and are replaced by a bewildering variety of social customs. In no other primate does

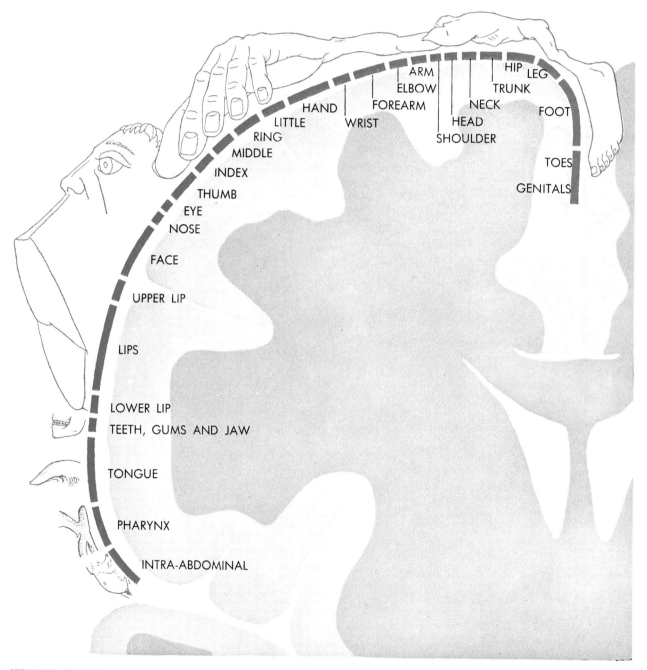

SENSORY HOMUNCULUS is a functional map of the sensory cortex of the human brain worked out by Wilder Penfield and his associates at the Montreal Neurological Institute. As in the map of the sensory cortex of the monkey that appears on the preceding page, the distorted anatomical drawing (*color*) indicates the areas of the sensory cortex associated with the various parts of the body.

a family exist that controls sexual activity by custom, that takes care of slow-growing young, and in which—as in the case of primitive human societies—the male and female provide different foods for the family members.

All these family functions are ultimately related to tools, hunting and the enlargement of the brain. Complex and technical society evolved from the sporadic tool-using of an ape, through the simple pebble tools of the man-ape and the complex toolmaking traditions of ancient men to the hugely complicated culture of modern man. Each behavioral stage was both cause and effect of biological change in bones and brain. These concomitant changes can be seen in the scanty fossil record and can be inferred from the study of the living forms.

Surely as more fossils are found these ideas will be tested. New techniques of investigation, from planned experiments in the behavior of lower primates to more refined methods of dating, will extract wholly new information from the past. It is my belief that, as these events come to pass, tool use will be found to have been a major factor, beginning with the initial differentiation of man and ape. In ourselves we see a structure, physiology and behavior that is the result of the fact that some populations of apes started to use tools a million years ago. The pebble tools constituted man's principal technical adaptation for a period at least 50 times as long as recorded history. As we contemplate man's present eminence, it is well to remember that, from the point of view of evolution, the events of the last 50,000 years occupy but a moment in time. Ancient man endured at least 10 times as long and the man-apes for an even longer time.

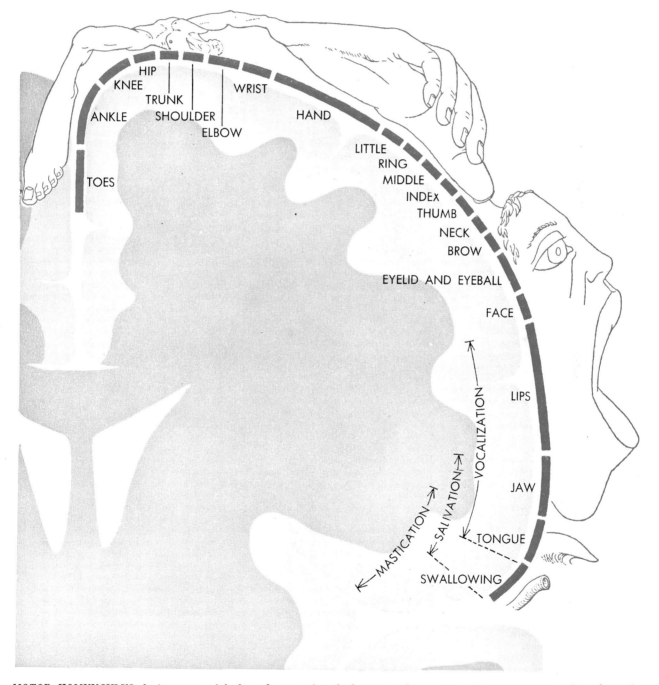

MOTOR HOMUNCULUS depicts parts of body and areas of motor cortex that control their functions. Human brain is shown here in coronal (ear-to-ear) cross section. Speech and hand areas of both motor and sensory cortex in man are proportionately much larger than corresponding areas in apes and monkeys, as can be seen by comparing homunculi with diagram of monkey cortex.

2

The Early Relatives of Man

by Elwyn L. Simons
July 1964

New evidence from fossils of the period between 60 and 12 million years ago not only illuminates the main stages of primate evolution but also singles out the ape stock from which the human line arose

A major feature of biological evolution during the past 70 million years has been the rapid rise to a position of dominance among the earth's land-dwelling vertebrates of the placental mammals (mammals other than marsupials such as the kangaroo and primitive egg-laying species such as the platypus). A major feature, in turn, of the evolution of the placental mammals has been the emergence of the primates: the mammalian order that includes man, the apes and monkeys. And a major event in the evolution of the primates was the appearance 12 million to 14 million years ago of animals, distinct from their ape contemporaries, that apparently gave rise to man.

Much of the evidence of the origin of man is new, but by no means all of it. For many years students of human evolution have broadly agreed that man's earliest ancestor would be found among the apelike primates that flourished during Miocene and early Pliocene times, roughly from 24 million to 12 million years ago [*see illustration on page 27*]. As long ago as the 1920's William K. Gregory of the American Museum of Natural History, after studying the limited number of jaw fragments and teeth then available, flatly pronounced man to be "a late Tertiary offshoot of the *Dryopithecus-Sivapithecus* group, or at least of apes that closely resembled those genera in the construction of jaw and dentition."

Until recently students of primate evolution have had little more evidence to work with than Gregory and his contemporaries did. Within the past 15 years, however, a number of significant new finds have been made—some of them in existing fossil collections. The early primates are now represented by many complete or nearly complete skulls, some nearly complete skeletons,

a number of limb bones and even the bones of hands and feet. In age these specimens extend across almost the entire Cenozoic era, from its beginning in the Paleocene epoch some 63 million years ago up to the Pliocene, which ended roughly two million years ago.

Sometimes a single jaw can tell a remarkably detailed evolutionary story, but there are no greater paleontological treasures than reasonably complete skulls and skeletons. Many such specimens have become available in recent years, but they do not lie in the exact line of man's ancestry. They are nonetheless important to the evolutionary history of all the primates. Both by their relative completeness and by their wide distribution in time they reveal new details concerning the main stages through which the primates probably passed during their evolutionary development.

To describe these stages is one of the two objectives of this article. The other objective is to summarize what is known about the relation of the early primates to the primate order's more advanced lineages, including man's own family: the Hominidae. The accomplishment of these objectives will show that the weight of today's knowledge fortifies Gregory's declaration of the 1920's.

Subdividing the Primates

Ideally zoological classification uses standard suffixes to guide the student through the maze of descending divisions: from class to order and thence —by way of suborders, infraorders, superfamilies, families, subfamilies and the like—to a particular genus and species. The grammar of primate taxonomy is not this simple. Two factors are responsible. First, there is no international agreement as to how the order of primates should be subdivided. Sec-

ond, generations of literary usage preceding man's first awareness of evolution have made all nouns derived from the Greek *anthropos* or the Latin *homo* virtually synonymous.

Nonetheless, an ability to read these taxonomic signposts is vital to an under

EARLY PRIMATE, about the size of a cat, was discovered in a Wyoming fossil deposit of middle Eocene age. One of the prosim-

standing not only of the relations among the 50-odd genera of living primates but also of the positions assigned to various extinct primates. This is because modern classification interrelates organisms in a pattern that reflects their evolutionary relations. In tracing the subdivisions that lead to man, for example, the first major branching divides the whole group of living primates into two suborders [*see illustration on the next two pages*]. The less advanced primates are assigned to the Prosimii; they are the various tree shrews, the many kinds of lemurs, the less abundant lorises and the solitary genus of tarsiers. The earliest known fossil primates belong exclusively to this suborder. The line to man, however, runs through successive divisions of the second primate suborder.

This suborder, consisting of the more advanced primates, is the Anthropoidea. It is divided into two infraorders. The less advanced anthropoids, including all the primates native to the New World, are the Platyrrhini. "Platyr-

rhine," which literally means "broad-nosed," refers to the wide spacing of the nostrils that is characteristic of the New World anthropoids.

The more advanced anthropoids are the Catarrhini. They include all other living anthropoids: the Old World monkeys, the apes and man himself. "Catarrhine" is opposed to "platyrrhine"; it literally means "hooknosed" but refers to a close spacing of the nostrils. The catarrhine infraorder is in turn divided into two superfamilies: the Cercopithecoidea and the Hominoidea. The first of these means "apes with tails"; it embraces the two subfamilies and 13 genera of living Old World monkeys.

The second catarrhine superfamily, the Hominoidea, embraces the subdivisions that finally separate the genus *Homo* from the rest of the living primates. The hominoids are split three ways: the families Hylobatidae, Pongidae and Hominidae. The first of these takes its name from *Hylobates*, the gibbon of South Asia, and includes both

this hominoid primate and the closely related siamang of Sumatra. The family Pongidae embraces the three genera of great apes: *Pongo*, the orangutan; *Pan*, the chimpanzee, and *Gorilla*, whose scientific name is the same as the common.

Of the family Hominidae, however complex its subfamilies and genera may or may not once have been, there survives today only the single genus *Homo* and its single species *Homo sapiens*. Man, then, is the sole living representative of the hominid family within the hominoid superfamily of the catarrhine infraorder of anthropoids. Or, to reverse the order of classification, among the 33 or so living genera of Anthropoidea whose names are accepted as valid there are only six genera of hominoids and a single hominid genus.

A Paleocene Tree-Dweller

The Age of Mammals was ushered in some 63 million years ago by a brief geological epoch: the Paleocene. Last-

ians, the less advanced of the two major divisions of the primate order, it is a member of the genus *Notharctus* and so belongs to a once abundant subfamily of tree-dwelling primates. Although primitive, the latter had many features in common with today's lemurs. They did not, however, give rise to any living prosimians and were extinct before the end of the Eocene, 36 million years ago.

ing perhaps five million years, the Paleocene was followed by the much longer Eocene epoch, which occupied roughly the next 22 million years. Both periods seem to have been characterized by warm temperatures that permitted tropical and subtropical forests to extend much farther north and south of the Equator than is the case today. These forests were inhabited by a diverse and abundant population of primates [see illustration on page 27]. The fossil record shows that species belonging to nearly 60 genera of prosimians, the bulk of them grouped in eight families, inhabited the Northern Hemisphere during Paleocene and Eocene times.

Three of these eight prosimian families are characterized by elongated front teeth, presumably adapted for chiseling and gnawing, as are the rather similar teeth of today's rodents and rabbits. It seems reasonable to suppose that these early primates started their evolutionary careers in competition for some kind of nibblers' and gnawers' niche in the warm forests. They were not successful; before the middle of the Eocene all three chisel-toothed prosimian families had become extinct. Perhaps they were put out of business by the rodents, which became abundant as these prosimians were dying out.

The skeletal remains of a member of one of these extinct families were recently found by D. E. Russell of the French national museum of natural history in late Paleocene strata near Cernay-lez-Reims in France. This early fossil primate belongs to the genus *Plesiadapis*, and the Cernay discovery includes a remarkably complete skull and a relatively complete series of limb and foot bones. An incomplete *Plesiadapis* skeleton is also known from Paleocene deposits in Colorado, and there are numerous jaws, jaw fragments and teeth from many other North American sites. These discoveries in opposite hemispheres, incidentally, make *Plesiadapis* the only genus of primate other than man's own that has inhabited both the Old and the New World.

Species of *Plesiadapis* varied in size from about the size of a squirrel to the size of a housecat. In life they probably looked as much like rodents as they did like primates [see illustration on page 28]. The patterns of the crowns of *Plesiadapis'* cheek teeth, however, resemble such patterns in lemur-like fossil primates of the Eocene epoch, and the structure of its limb bones links it with such living prosimians as the lemurs of the island of Madagascar (now the Malagasy Republic).

Plesiadapis is nonetheless distinctive. Its skull has a small braincase and a long snout. Its enlarged and forward-slanting incisors are widely separated from its cheek teeth. This arrangement is characteristic of the rodents, and although *Plesiadapis* appears too late to be an ancestor of the rodents, some workers have suggested that the order of rodents may be descended from animals not very different from it.

Plesiadapis exhibits two other traits that set the genus apart from almost all later primates. First, most if not all of its fingers and toes ended in long claws that were flattened at the sides. Among living primates only the tree shrews have a claw on each digit; all other species have either a combination of nails and claws or nails exclusively. Moreover, the claws of living primates are small compared with those of *Plesiadapis*. Regardless of their size, these claws probably served the same function as claws do among living tree shrews, helping this ancient arboreal primate to scramble up and down the trunks of trees.

The second peculiar trait, possibly one of lesser significance, is a resemblance between the structure of the middle ear of *Plesiadapis* and that of a nonprimate: the colugo, or "flying lemur," which still inhabits southeast Asia. The first thing to be said of the colugos, as George Gaylord Simpson has put it, is that they "are not lemurs and cannot fly." Colugos are so unusual that taxonomists have been obliged to place them in a mammalian order—the Dermoptera—all their own. The size of a squirrel or larger, with broad flaps of skin for gliding that run from its forelimbs to the tip of its tail, the colugo shows little outward resemblance to any other living mammal. It has been conjectured that the colugos are ultimately

TAXONOMY of the living primates ranks the order's 52 genera (*scientific and common names at far right*) according to divisions of higher grade. There is no universal agreement on how this should be done. For example, the two infraorders of anthropoids in this system are held by many investigators to be suborders and thus equal in rank with the Prosimii. It is generally agreed, however, that man belongs among the catarrhines and, within that group, is a member of the hominoid superfamily (as are all the apes) and the hominid family, in which he is the only living species of the genus *Homo*.

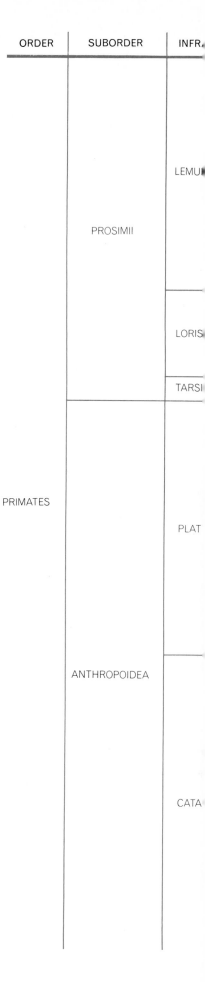

ORDER	SUBORDER	INFR
	PROSIMII	LEMU
		LORIS
		TARSI
PRIMATES		
	ANTHROPOIDEA	PLAT
		CATA

SUPERFAMILY	FAMILY	SUBFAMILY	GENUS	COMMON NAME
TUPAIOIDEA	TUPAIIDAE	TUPAIINAE	TUPAIA DENDROGALE UROGALE	COMMON TREE SHREW SMOOTH-TAILED TREE SHREW PHILIPPINE TREE SHREW
		PTILOCERCINAE	PTILOCERCUS	PEN-TAILED TREE SHREW
LEMUROIDEA	LEMURIDAE	LEMURINAE	LEMUR HAPALEMUR LEPILEMUR	COMMON LEMUR GENTLE LEMUR SPORTIVE LEMUR
		CHEIROGALEINAE	CHEIROGALEUS MICROCEBUS	MOUSE LEMUR DWARF LEMUR
	INDRIDAE		INDRI LICHANOTUS PROPITHECUS	INDRIS AVAHI SIFAKA
	DAUBENTONIIDAE		DAUBENTONIA	AYE-AYE
LORISOIDEA	LORISIDAE		LORIS NYCTICEBUS ARCTOCEBUS PERODICTICUS	SLENDER LORIS SLOW LORIS ANGWANTIBO POTTO
	GALAGIDAE		GALAGO	BUSH BABY
TARSIOIDEA	TARSIIDAE		TARSIUS	TARSIER
CEBOIDEA	CALLITHRICIDAE		CALLITHRIX LEONTOCEBUS	PLUMED AND PYGMY MARMOSETS TAMARIN
	CEBIDAE	CALLIMICONINAE	CALLIMICO	GOELDI'S MARMOSET
		AOTINAE	AOTES CALLICEBUS	DOUROUCOULI TITI
		PITHECINAE	PITHECIA CHIROPOTES CACAJAO	SAKI SAKI UAKARI
		ALOUATTINAE	ALOUATTA	HOWLER
		CEBINAE	CEBUS SAIMIRI	CAPUCHIN SQUIRREL MONKEY
		ATELINAE	ATELES BRACHYTELES LAGOTHRIX	SPIDER MONKEY WOOLLY SPIDER MONKEY WOOLLY MONKEY
CERCOPITHECOIDEA	CERCOPITHECIDAE	CERCOPITHECINAE	MACACA CYNOPITHECUS CERCOCEBUS PAPIO THEROPITHECUS CERCOPITHECUS ERYTHROCEBUS	MACAQUE BLACK APE MANGABEY BABOON DRILL GELADA GUENON PATAS MONKEY
		COLOBINAE	PRESBYTIS PYGATHRIX RHINOPITHECUS SIMIAS NASALIS COLOBUS	COMMON LANGUR DOUC LANGUR SNUB-NOSED LANGUR PAGI ISLAND LANGUR PROBOSCIS MONKEY GUERAZA
HOMINOIDEA	HYLOBATIDAE		HYLOBATES SYMPHALANGUS	GIBBON SIAMANG
	PONGIDAE		PONGO PAN GORILLA	ORANGUTAN CHIMPANZEE GORILLA
	HOMINIDAE		HOMO	MAN

related to both the primates and the bats. The resemblance in ear structure is not the only similarity between the living colugos and the long-extinct *Plesiadapis:* the colugo's digits also bear sizable claws. Both of these similarities, however, could have been acquired independently rather than from a common ancestor.

Although early in time and cosmopolitan in range, *Plesiadapis* is clearly too specialized a primate to be the ancestor of later prosimians. This sterile offshoot of the family tree is significant to primate history on other grounds. First, the relative completeness of its remains makes *Plesiadapis* the most thoroughly known primate of the Paleocene. Second, many details of its skeletal form serve to link its order with that of the even earlier placental mammals—the Insectivora, from which the primates arose.

Eocene Evolutionary Advances

The next fossil primates of which there are nearly complete remains come from North American strata of the middle Eocene. The best-known examples are species of two related lemur-like genera: *Notharctus* and *Smilodectes.* The degree to which these prosimians have advanced beyond *Plesiadapis* demonstrates the rapid evolution of primates as much as 50 million years ago. Many incomplete specimens of *Notharctus* were exhaustively studied in the 1920's by Gregory. Since then an even more complete skeleton of one species —probably *Notharctus tenebrosus*—has come to light in the paleontological research collection of Yale University. Although the skull is missing, the rest of the skeleton represents one of the two most complete individual primates yet recovered from fossil beds of such early date. C. Lewis Gazin of the Smithsonian Institution has recently recovered several complete skulls and many other bones of *Smilodectes gracilis* in southwestern Wyoming. The abundance of this new material has permitted the assembly of a skeleton and a restoration of *Smilodectes'* probable appearance [*see illustration on page 29*].

These New World primates resemble living lemurs both in their proportions and in their general structure. In contrast to the small-brained, snouty, side-eyed *Plesiadapis,* the skull of *Smilodectes* shows an enlargement of the front portion of the brain and a shifting of eye positions forward so that individual fields of vision can overlap in

front. These features of the head, taken together with the animal's rather long hind limbs, suggest that in life *Smilodectes* looked rather like one of today's Malagasy lemurs, the sifaka.

It is most unlikely, however, that either *Smilodectes* or *Notharctus* contributed to the ancestry of living lemurs. This honor can more probably be conferred on some member of a European genus, such as *Protoadapis* or *Adapis,* of equal Eocene age, if indeed the ancestors of modern lemurs were not already in Africa by this time. *Adapis* has the distinction of being the first fossil primate genus ever described. The French paleontologist Baron Cuvier did so in 1822, although he originally thought *Adapis* was a hoofed mammal or a small pachyderm and not a primate at all. Unfortunately none of these possible Old World precursors of living lemurs is sufficiently represented by fossils to provide the kind of detailed skeletal information we possess for their New World contemporaries.

This is also the case for a roughly contemporary European prosimian: *Necrolemur,* known from skulls and limb bones found in the Quercy deposits of France and by extrapolation from parts of a related species recovered in Germany. In *Necrolemur* the evolutionary advances represented by *Notharctus* and *Smilodectes* have been extended. Enlargement of the forebrain and a further facial foreshortening are apparent. A forward shift of the eye position —with the consequent overlapping of visual fields and potential for depth perception—should have equipped *Necrolemur* for an active arboreal life in the Eocene forests. Actually this early primate, although it is probably not ancestral to any living prosimian, shows a much closer affinity for the comparatively advanced tarsier of southeast Asia than for the more primitive Malagasy lemurs.

The evolutionary progress made by prosimians during the Eocene, both in North America and in Europe, is obvious. Yet not a single fossil primate of the Eocene epoch from either continent appears to be an acceptable ancestor for the great infraorder of the catarrhines, embracing all the living higher Old World primates, man included. One cannot help wondering what developments may have been taking place in Africa and Asia during the Eocene's span of more than 22 million years. In both regions the record is almost mute. In Asia the only known primate fossils dating to this epoch are a few equivocal bits and pieces from China and some

fragments from a late Eocene formation in Burma. From the Eocene of Africa there are not only no primates but also no small mammals of any kind.

One of the Burmese fragments is a section of lower jaw containing three premolar teeth and one molar, described in 1938 by Edwin H. Colbert of the American Museum of Natural History, who named the new species *Amphipithecus mogaungensis.* A brief lesson in primate teeth is necessary to understand its significance. The lesson is painless; it merely involves counting. The facts are these: Regardless of tooth size or shape all adult catarrhines—Old World monkeys, apes and man—have the same "dental formula." In each half of a jaw—upper and lower alike—are found from front to back two incisors, a single canine, two premolars and three molars. In anatomical shorthand the fact is written:

$$\frac{2:1:2:3}{2:1:2:3} \times 2 = 32 \, .$$

Because of its three premolar teeth *Amphipithecus* is dentally more primitive than any catarrhine, fossil or living. It may have had such a dental formula as

$$\frac{2:1:3:3}{2:1:3:3} \times 2 \, .$$

This is typical of some living lemurs and of many platyrrhines—the marmosets and monkeys of the New World. Yet in other characteristics the *Amphipithecus* jaw is advanced rather than primitive. The horizontal ramus—that portion of the jaw that holds the teeth— is deep and massive, as is also true in many fossil and living apes. The fossil premolars, and the molar as well, are

PHYLOGENY of all the primates traces the evolution of the order from its beginnings sometime before the middle of the Paleocene (*see time scale in illustration at right*). The first to appear were prosimian families that stemmed from a basic stock of small and sometimes arboreal mammals called Insectivora (whose living kin include the shrews and moles). The chart's broken lines show hypothetical evolutionary relations. In the interval between Eocene and Miocene times these relations are particularly uncertain. Solid lines (*color*) show the periods when species of the groups named (*black type*) are known to have flourished. The names of two prosimian and two anthropoid genera appear in color. Species of each are illustrated in detail on the four following pages.

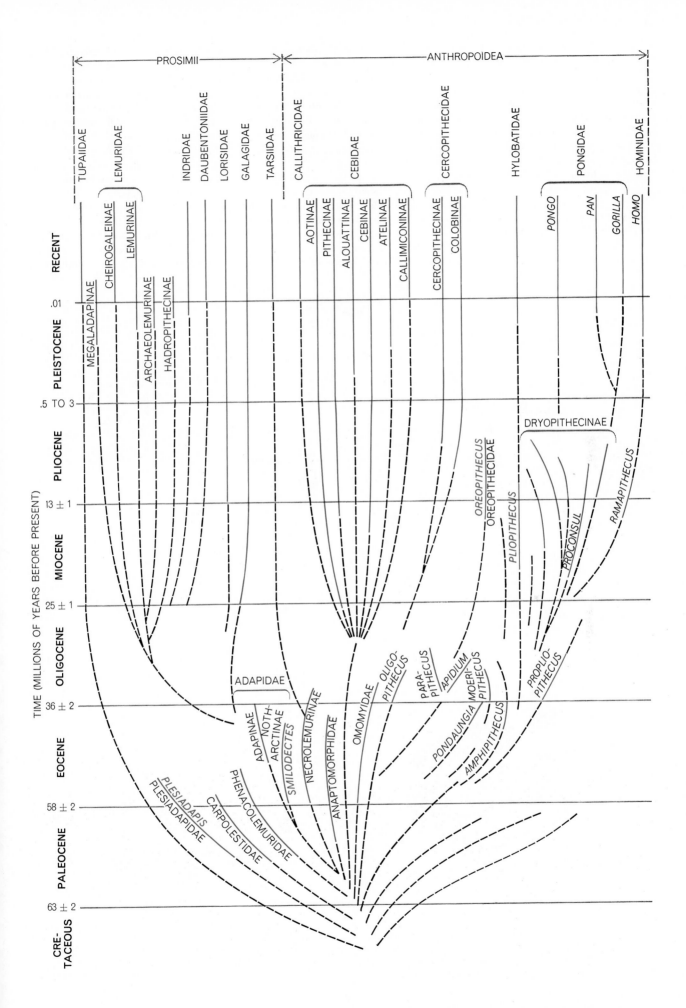

similar to the corresponding teeth in *Oligopithecus*, a newly discovered catarrhine from the Oligocene of Egypt.

The other Burmese fossil consists of both rear portions of a lower jaw, discovered together with a segment of upper jaw containing two molar teeth. G. E. Pilgrim of the Indian Geological Survey gave this find the name *Pondaungia cotteri* in 1927. The two molars almost equally resemble those of prosimians on the one hand and of some Old World higher primates on the other. The material is so fragmentary, however, that some scholars have even questioned *Pondaungia*'s inclusion in the primate order. If neither *Amphipithecus* nor *Pondaungia* were known, it would seem almost certain that the Old World anthropoids had arisen in Africa. Further collecting in the Burmese Eocene formation that contained both Pilgrim's and Colbert's fossils is required before final judgment of their significance can be made.

By the end of the Eocene the primates had been differentiating for almost 30 million years. This is a long time. Yet only one result is known with certainty: A number of primates, lemur-like and tarsier-like, had evolved in the Old World, some of which must have contributed to the ancestry of today's lower primates, the Prosimii. Not until the close of the Eocene do some puzzling fossil fragments from Burma offer a hint of what must have been a major, even though still undocumented, evolutionary development in the Old World Tropics. This development can be postulated with confidence, in spite of a paucity of evidence, because early in the following epoch—the Oligocene—fossil Anthropoidea appear in substantial numbers and varieties. It is highly improbable that these Oligocene primates could have evolved, in terms of geologic time, almost overnight. So

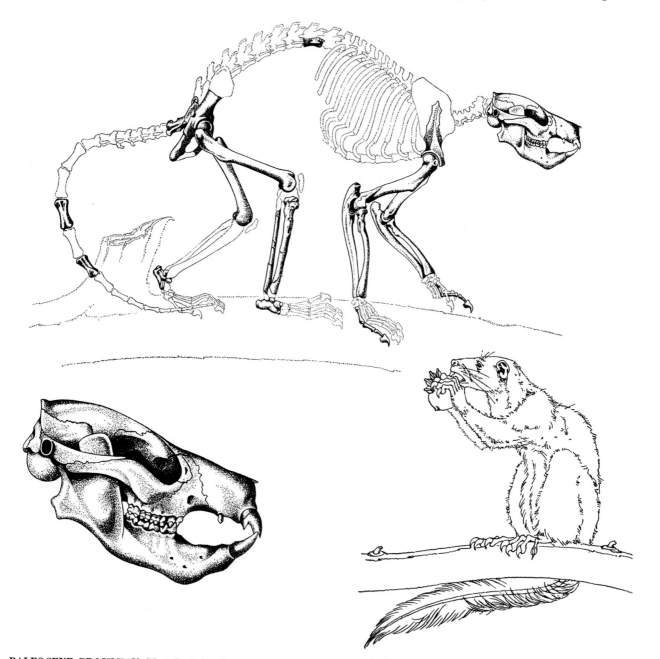

PALEOCENE PROSIMIAN *Plesiadapis* has been reconstructed (*skeleton at top*) on the basis of French and North American fossil finds and restored (*figure at bottom right*) by analogy with living tree shrews. Bones shown in outline are hypothetical. The charac- teristic wide gap between this rodent-like animal's cheek teeth and its slanting incisors is evident in the skull detail (*bottom left*). Species of *Plesiadapis* ranged from squirrel- to cat-size. They belong to an early primate family that died out 50 million years ago.

far our knowledge of their geographical distribution is exceedingly limited: all their remains discovered to date have come from a single formation in the desert badlands of the Egyptian province of the Fayum.

The Catarrhine Emergence

A hundred miles inland from the Mediterranean coast and some 60 miles southwest of Cairo a brackish lake stands at the edge of a series of escarpments and desert benches that are almost devoid of plant and animal life. At the end of the Eocene epoch the shore of the Mediterranean extended this far inland, and rivers flowed into the shallow sea through dense tropical forests. The rise and fall of sea and land is clearly revealed by alternating river-deposited strata and layers of marine limestone. In the middle of these escarpments, running from southwest to northeast between the lake and a lava-capped ridge called Gebel el Quatrani, is a fossil-rich stratum of sandy early Oligocene sediments that first yielded primate remains in the early 1900's.

Primates were not the only inhabitants of this forested Oligocene shoreline. Crocodiles and gavials swam in the sluggish streams. Tiny rodents and various relatives of today's hyrax lived in the underbrush, as did hog- and ox-sized cousins of the modern elephant. The largest animal of the fauna was a four-horned herbivore about the size and shape of today's white rhinoceros.

Until the recent Yale Paleontological Expedition the primate inventory from the Fayum totaled seven pieces of fossilized bone: one skull fragment (picked up by a professional collector in 1908 and sent to the American Museum of Natural History), one heel bone, three fragmentary portions of jawbone and two nearly complete lower jaws. This

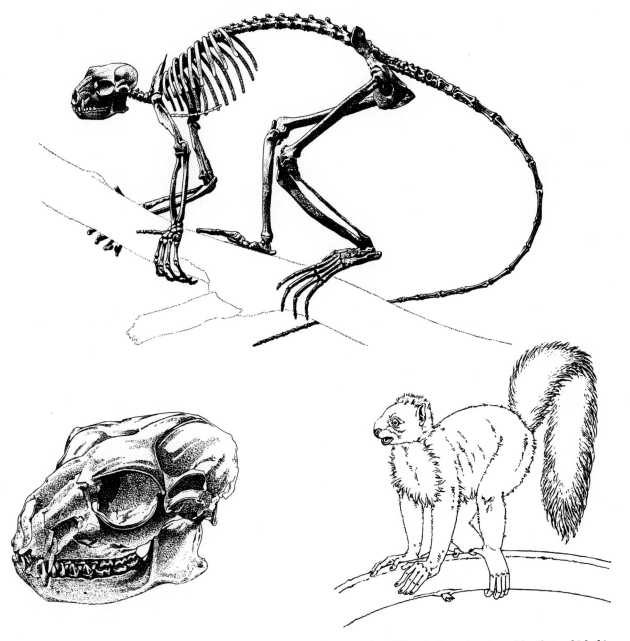

EOCENE PROSIMIAN *Smilodectes* is several million years junior to *Plesiadapis* and is a far more advanced animal. Its snout is shorter, the front portion of the brain is enlarged and its eyes are positioned on the skull in a manner that permits the visual fields to overlap. Although the notharctine subfamily to which this genus belongs was not ancestral to any of the living primates, its relatively long hind limbs give *Smilodectes* a remarkable resemblance to one modern prosimian, the sifaka, a Malagasy lemur.

may not seem a particularly rich haul, but studies over the years have shown that these seven fossils represent at least four distinct genera and species of Oligocene primates.

By the end of the Yale Expedition's fourth season this past winter more than 100 individual primate specimens had been added to the Fayum inventory. Although many of these finds consist of single teeth, there are also more than two dozen lower jaws, a skull fragment and some limb bones. Thus far the Fayum beds have not yielded any skulls or other skeletal remains of the kind that provide so much detailed information on Paleocene and Eocene prosimians. What has been found, however, reveals a great deal. As one example, an incomplete lower jaw was discovered in 1961 by a member of the expedition, Donald E. Savage of the University of California at Berkeley. This fragment permits the establishment of a new primate genus, which I have named *Oligopithecus*. The molar teeth of the "type" species of the genus indicate that it may well be on or near the evolutionary line that gave rise to the superfamily of living Old World monkeys: the cercopithecoids.

The other Old World primate superfamily—the hominoids—also appears to be well represented among the Fayum fossils. Possible ancestors for one family of living hominoids—the gibbons and siamangs—are present: the well-preserved jaw of a gibbon-like animal, as yet undescribed, was turned up by the Yale Expedition in 1963. In this connection it should be noted that the study of all the Fayum fossils belonging to the genus *Propliopithecus*—for many years regarded as an ancestor of the gibbon—indicates that it probably represents a more generalized hominoid ancestor instead. This small Oligocene primate may well prove to be on or

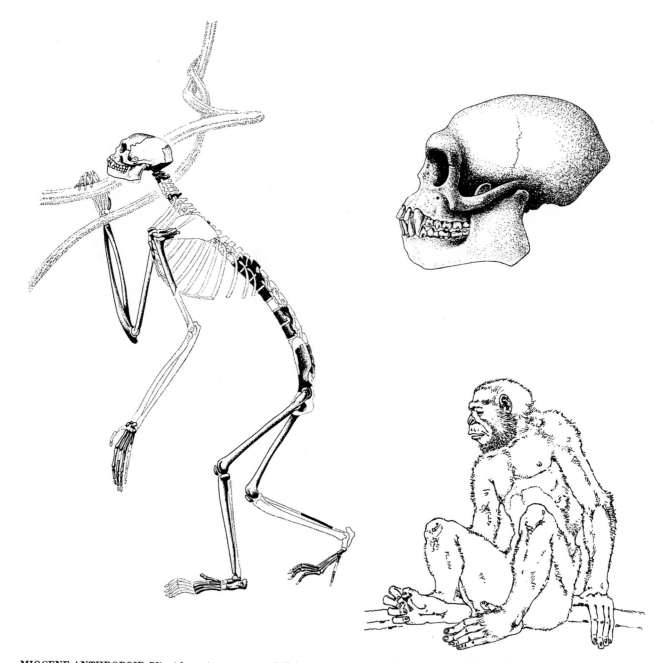

MIOCENE ANTHROPOID *Pliopithecus* is reconstructed (*left*) on the basis of a fairly complete specimen discovered in 1957. Although it is as much as 20 million years old, its skull (*top right*) is very much like a modern gibbon's. These hominoids probably were ancestral to today's long-armed gibbons, but their anatomy is generalized and their fore- and hind limbs are of almost equal length.

near the line of evolutionary development that led to the living pongids and to man.

The Miocene Hominoids

Throughout the entire 11-million-year span of the Oligocene the fossil fauna of Europe does not include a single primate. In the following epoch—the Miocene, which had its beginning some 24 to 26 million years ago—primates reappear in the European fossil record. A few years after Cuvier named *Adapis* the paleontologist-antiquarian Édouard Lartet reported a primate low-er jaw from Miocene strata at Sansan in France. This fossil was the basis for establishing the genus *Pliopithecus*. Since then dozens of other *Pliopithecus* specimens have been uncovered in formations of Miocene and Pliocene age, in both Europe and Africa. The best of these *Pliopithecus* finds to date—a skull, including facial portions, and most of a skeleton—was made in a Miocene deposit near the Czechoslovakian town of Neudorf an der March in 1957. These remains provided the basis for the reconstruction shown on the opposite page.

Many millions of years younger than the gibbon-like hominoids of the Fa-yum, *Pliopithecus* presumably represents a further advance in the lineage that leads to the living gibbons. Yet this Miocene hominoid shows quite generalized characteristics. The arms of today's gibbons are considerably longer than their legs; *Pliopithecus*, in contrast, has hind limbs and forelimbs of nearly equal length. In fact, where comparisons are possible, *Pliopithecus* is not radically different from other roughly contemporary but not as fully preserved Miocene hominoids. Study of its skeleton tells us much about what the early hominoids were like.

A near contemporary of *Pliopithecus*

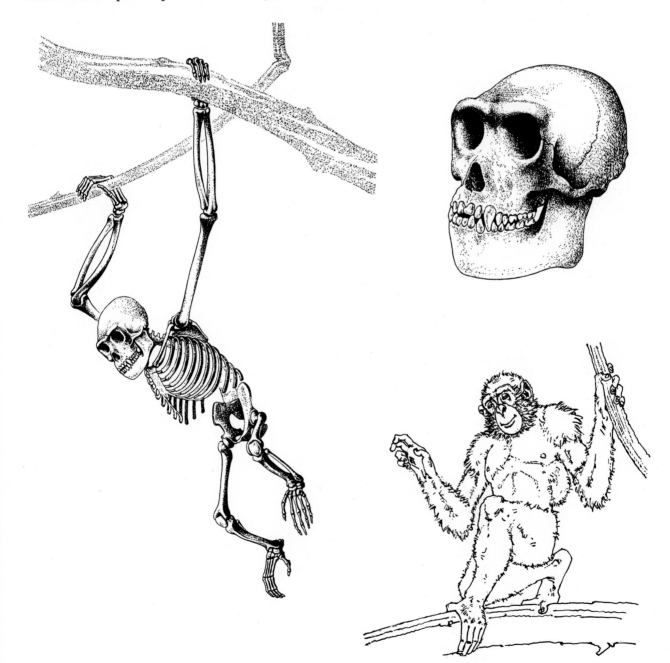

PLIOCENE ANTHROPOID *Oreopithecus* is not more than 14 million years old. Because most fossil anthropoids are small and their faces are snouty, the skeleton on which this reconstruction is based caused a sensation when first discovered. Its flat profile, four-foot height and also an apparent ability to walk erect brought *Oreopithecus* passing notoriety as a possible hominid "missing link."

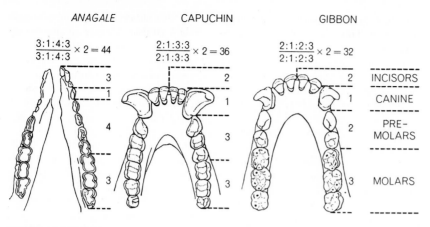

ANAGALE   CAPUCHIN   GIBBON

$$\frac{3:1:4:3}{3:1:4:3} \times 2 = 44 \qquad \frac{2:1:3:3}{2:1:3:3} \times 2 = 36 \qquad \frac{2:1:2:3}{2:1:2:3} \times 2 = 32$$

3	2	2	INCISORS
1	1	1	CANINE
4	3	2	PRE-MOLARS
3	3	3	MOLARS

HIGHER PRIMATES have fewer teeth than the original placental total of 44 (*see the lower jaw of* Anagale, *left*). The platyrrhine primates have lost an incisor and a premolar on each side of both jaws, and some have even lost molars. Thus the New World cebids have 36 teeth (*capuchin monkey, center*). All the catarrhines have lost one more premolar all around, so the Old World monkeys, apes and man have only 32 teeth (*gibbon, right*).

is *Dryopithecus*, the animal mentioned by Gregory as one candidate for a position ancestral to man. *Dryopithecus* was also named by Lartet; he described a lower jaw in 1856, almost 20 years after his discovery of *Pliopithecus*. Since that time many other fossil fragments of *Dryopithecus*—but no complete skulls or skeletons—have been found in strata of Miocene and even Pliocene age in Europe. In the late 1950's fossil teeth assignable to *Dryopithecus* were uncovered in brown coal deposits in southwestern China, indicating that the range of these hominoids extended across Europe to the Far East.

Because the fossil inventory for *Dryopithecus* consists mainly of individual teeth and teeth in incomplete jawbones, the reader will find useful some additional facts about primate dentition. These facts concern shape rather than number. First, although the crowns, or chewing surfaces, of any primate's molars may be ground flat by years of wear, each crown normally shows several bumps called cusps. Typically there are four cusps to a crown, one at each corner of the tooth. Second, all members of one of the two higher Old World primate superfamilies—the cercopithecoid monkeys—exhibit a unique cusp pattern. On the first and second upper and lower molars ridges of enamel project toward each other from the front pair of cusps; there are similar ridges between the back pair of cusps. Before the crown has been worn down there is often a gap in the middle of the ridge, but worn or unworn these molars are unmistakable.

The hominoids, on the other hand, have their own distinctive cusp pattern. The lower molars normally have five cusps rather than four, and the pattern of valleys that separate these bumps of enamel somewhat resembles the let-ter Y, with the bottom of the Y facing forward. The pattern is called Y-5. The evolutionary significance of the pattern lies in the fact that the lower molars of *Dryopithecus* and of early men typically exhibit a Y-5 pattern. Thus Y-5 is a hereditary characteristic that has persisted among the hominoids for at least 24 million years.

Because the *Dryopithecus* fossil remains in Europe and the Far East are fragmentary, they reveal almost nothing about the skull and skeleton of this hominoid. Fortunately discoveries in Africa have altered the situation. There, thanks to the untiring efforts of L. S. B. Leakey and his colleagues, a substantial inventory of Miocene primate fossil remains has been accumulated, most of them from Rusinga Island in Lake Victoria and the nearby shores of the lake. As a result several species of African proto-ape, apparently ranging from the size of a gibbon to that of a gorilla, have been described.

In spite of this variety in size, all these species are assigned to the single genus *Proconsul*. The name *Proconsul* is an "inside" British joke: the "pro" is simply "before" but "Consul" was the pet name of a chimpanzee that had long been a beloved resident of the London Zoo. All jokes apart, the name implies an evolutionary position for these African hominoids close to the ancestry of living chimpanzees and quite possibly to the ancestry of gorillas as well.

Among the *Proconsul* species the fossil remains that are most nearly complete belong to the gibbon-sized *Proconsul africanus*. They include parts of two skulls—almost complete in the facial portions—and some limb bones, including parts of a foot and a forelimb with a hand. The picture that emerges from the study of this material is that of an advanced catarrhine, showing some monkey-like traits of hand, skull and brain but hominoid and even partially hominid characteristics of face, jaws and dentition. The foot and forelimb are also more suggestive of some ape-like adaptations—including an incipient ability to swing by the arms from tree branch to tree branch—than they are of either arboreal or ground-dwelling Old World monkeys.

Recent taxonomic investigations show that species of the genus *Proconsul*, with their relative abundance of skeletal remains, should almost certainly be lumped together with the genus *Dryopithecus*. What such an assignment would mean, in effect, is that all these Miocene-Pliocene hominoids—not only Eurasian but African as well—belong to

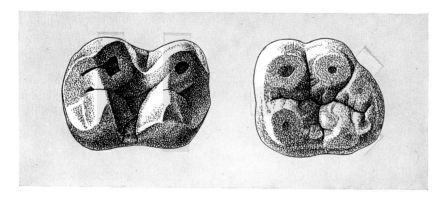

SHAPE OF MOLAR TEETH serves to split the catarrhines into two groups. The crowns of Old World monkeys' molars have a cusp at each corner (*baboon, left*): both front and rear pairs of cusps are connected by ridges (*color*). The crowns of apes' and man's lower molars normally have five cusps (*chimpanzee, right*), and the "valleys" between the cusps resemble the letter Y (*color*). This Y-5 pattern first appeared some 24 million years ago.

a single cosmopolitan genus. This might have been recognized 30 years ago except for a series of mischances. A. T. Hopwood of the British Museum (Natural History), who named *Proconsul* in 1933, stated that the lower jaws and teeth of *Proconsul* and *Dryopithecus* could not be distinguished as belonging to separate genera. He found the opposite to be true of the upper teeth, but it happens that the particular specimen of *Dryopithecus* upper teeth he chose for comparison was not of that genus at all: it belonged to another primate, *Ramapithecus*. When W. E. Le Gros Clark of the University of Oxford and Leakey later enlarged the definition of *Proconsul*, they still drew the primary upper-dental distinctions from the same specimen, which was not recognized as *Ramapithecus* until 1963. *Proconsul* and *Ramapithecus* are not the same genus. *Proconsul* and *Dryopithecus*, in all probability, are.

The Puzzle of the Coal Man

In any listing of the more complete early primates the Italian species *Oreopithecus bambolii*, sometimes irreverently known as the Abominable Coal Man, cannot be omitted. Its first bits and pieces were discovered almost 100 years ago. Since then *Oreopithecus* remains have been found in abundance in the brown-coal beds of central Italy, a formation that is variously assigned to late Miocene or early Pliocene times. In 1956 Johannes Hürzeler of the natural history museum in Basel assembled a number of new *Oreopithecus* specimens, and in 1958 Hürzeler was instrumental in the recovery of a nearly complete skeleton from a coal mine at Grosseto in Italy. This superb fossil is still being examined by specialists from various nations.

Evidently these Miocene-Pliocene primates were of substantial size—some four feet tall and probably weighing more than 80 pounds. Among the living primates the closest in size would be a female chimpanzee. Because its face is short and flat instead of showing an elongated snout, and because studies of its pelvis and limb bones suggest the possibility of an erect walking posture, *Oreopithecus* has received some notoriety as a possible direct precursor of the hominid family. Intensive study of the 1958 specimen, however, has led a number of workers to rather different conclusions.

One of the surprising things about *Oreopithecus* was first noted by Gregory in the 1920's: the cheek teeth of its lower jaw strongly resemble the corresponding teeth of *Apidium*, one of the four primates named from the original Fayum finds of the early 1900's. The surprise is that *Apidium* dates to the Oligocene, some 20 to 25 million years earlier than *Oreopithecus*. This remarkable dental coincidence might easily have remained no more than a curiosity if the Yale Expedition had not recovered a number of additional *Apidium* teeth— this time the cheek teeth from upper jaws. The study of these teeth is not yet complete, but it is already evident that the newfound *Apidium* uppers correspond as well to the equivalent *Oreopithecus* uppers as the lowers do to the lowers. Such a similarity strongly suggests that, in spite of their separation in time, the ancient *Apidium* and the comparatively modern *Oreopithecus* are representatives of a single group of now extinct Old World higher primates. *Apidium* cannot be directly ancestral to *Oreopithecus*, however, because it lacks one pair of incisors that are still present in *Oreopithecus*. Although in the evolutionary sense this group is not far removed from the pongid-hominid stem, it seems to have developed its own distinctive characteristics by early Oligocene times.

A Dryopithecine from India

Having come to the end of Miocene times, with a scant 12 million to 14 million years remaining in which to discover a human forebear, we must reexamine Gregory's declaration. One of his candidates, *Dryopithecus*, has now been shown to be a long-lived and cosmopolitan genus, one of an abundant dryopithecine group to which in all probability the African species of *Pro-*

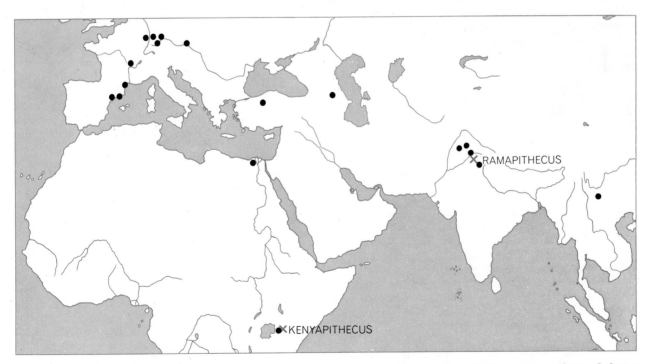

RANGE OF THE DRYOPITHECINES during Miocene and early Pliocene times extended across Eurasia from France to western China and also included East Africa and northwest India. No other hominoid primates of that time were so widespread. Crosses (*color*) show where the advanced hominoid genus, *Ramapithecus*, and the apparently identical *Kenyapithecus* have been found.

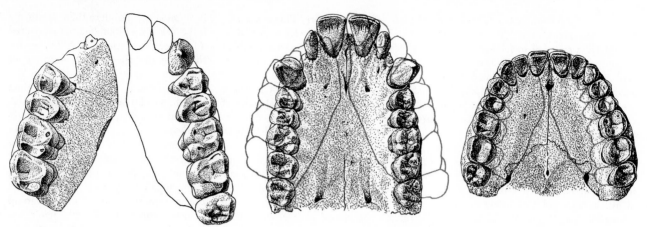

UPPER JAW OF RAMAPITHECUS, in a life-sized reconstruction at left, is compared with that of an orangutan (*center*) and a man (*right*). In each comparison the jaws have been made the same size. The U-shaped arc formed by the ape's teeth contrasts sharply with the curved arc in *Ramapithecus*, which is closer to the human curve. This, as well as the manlike ratio in comparative size of front and cheek teeth and the modest canine, are grounds for considering *Ramapithecus* man's earliest known hominid ancestor.

consul belong. What about *Sivapithecus*, Gregory's other nominee for a position as a hominid ancestor?

The Siwalik Hills of northwestern India and adjacent Pakistan have been known to paleontologists since the first half of the 19th century for their fossil-rich deposits of Miocene and Pliocene age. It was from these strata that Pilgrim, who described the Burmese borderline primate *Pondaungia*, uncovered and named *Sivapithecus* in 1910. Later, in the 1930's, G. Edward Lewis collected fossils for the Yale-Cambridge North India Expedition from these same beds and discovered a number of primate jaw fragments and teeth. In due course they were assigned to several separate primate genera, including some additional examples of Pilgrim's *Sivapithecus*.

Recent reexamination of *Sivapithecus* species suggests that they are not markedly different from *Dryopithecus*. Like *Proconsul*, they may well deserve nothing more than subgeneric status among the cosmopolitan dryopithecines. This would mean that not only Africa and Eurasia but also India supported separate populations of a single hominoid genus during Miocene and the earliest of Pliocene times—a span of at least 15 million years. However confused and confusing dryopithecine taxonomy and evolutionary relations are at present, the inescapable fact remains that throughout this entire span of time this is the only group of primates known in any Old World continent that can be considered close to the source of the hominid family line.

Because of the dryopithecines' very broad distribution throughout the Old World, the precise time and location of the primates' evolutionary advance from hominoid animals to specifically hominid ones may always remain uncertain. Yet a tentative guess is possible.

Another of the fossil primates Lewis collected in the Siwalik Hills was *Ramapithecus*. The type species of *Ramapithecus* is founded on a portion of a right upper jaw and is named *Ramapithecus brevirostris*. The fossil includes the first two molars, both premolars, the socket of the canine tooth, the root of the lateral incisor and the socket of the central incisor. When it and other fossils of *Ramapithecus* are used to reconstruct an entire upper jaw, complete with palate, the result is surprisingly human in appearance [*see illustration above*]. The proportions of the jaw indicate a foreshortened face. The size ratio between front teeth and cheek teeth is about the same as it is in man. (The front teeth of living apes are relatively large.) Estimating from the size of its socket, the canine tooth was not much larger than the first premolar—another hominid characteristic, opposed to the enlarged canines of the pongids. The arc formed by the teeth is curved as in man, rather than being parabolic, or U-shaped, as in the apes.

From Relative to Ancestor

Just such traits as these, intermediate between the dryopithecines and hominids, had led Lewis in 1934 to suggest that *Ramapithecus* might well belong to the Hominidae. This suggestion was challenged in the years immediately following. In my opinion, however, both the reexamination of the type species and the identification of new material reinforce Lewis' original conclusion.

Taxonomic decisions of this sort are not made lightly, and to draw a large conclusion from limited fossil evidence is always uncomfortable. Thus it was particularly gratifying to learn of Leakey's recovery, in 1962, of the jaws of a similar hominid near Fort Ternan in southwestern Kenya.

Leakey has assigned this fossil to the species *Kenyapithecus wickeri*. Like the remains of Lewis' *Ramapithecus brevirostris*, it preserves much of the upper dentition. Included are the first two molars on both sides, an intact second premolar and the stub of a first premolar. The socket for one canine is intact; a canine tooth and a central incisor were found separately. Potassium-argon dating of the specimen yields an absolute age of about 14 million years, a time near the boundary between the Miocene and the Pliocene.

The significance of the Fort Ternan find lies in the fact that *Kenyapithecus* not only has an abundance of close anatomical links with *Ramapithecus* but also exhibits no pertinent differences. In this new specimen, a continent removed in space from *Ramapithecus*, are found the same foreshortened face, dental curve and small canine tooth—each a hominid trait. The conclusion is now almost inescapable: in late Miocene to early Pliocene times both in Africa and India an advanced hominoid species was differentiating from more conservative pongid stock and developing important hominid characteristics in the process. Pending additional discoveries it may be wiser not to insist that the transition from ape to man is now being documented from the fossil record, but this certainly seems to be a strong possibility.

Ramapithecus

by Elwyn L. Simons
May 1977

*This extinct primate is the earliest hominid, or
distinctively manlike, member of man's family tree.
The finding of many new specimens of it has clarified
its place in human evolution*

Fifteen years ago the only known evidence bearing on when and how the distinctively manlike branch of man's family tree first arose was a single fragment of upper jaw that had been found in northern India in 1932. Today numerous other fossils of this earliest-known hominid and genera closely related to it are known from East Africa, Greece, Turkey, Hungary, India and Pakistan. In the early 1960's the count of these hominid fossils, most of which were turning up not in the field but in museum collections of primate fossils around the world, increased from the single India specimen to nearly a score. Since 1972 that number has at least doubled. For primate paleontology it has been a remarkable progression from rags to riches. For the student of human evolution the new finds make it possible to more clearly discern an evolutionary pathway traversing the past 14 million years. That pathway can now be traced with little fear of contradiction from generalized hominoids (the larger branch of man's family tree that includes the apes) to the hominids and from the hominids to the genus *Homo*.

The pathway begins in Miocene times with an Old World population of apes whose existence became known more than a century ago. In 1856, Édouard Lartet, a French lawyer and paleontologist, reported on a primate jaw found in a clay of Miocene age at Saint Gaudens in the French Pyrenees. Lartet named the fossil species *Dryopithecus fontani*. The generic name, a combination of the Greek for "oak" and "ape," reflected Lartet's belief that the primate had lived in the forest. The animal and plant remains found in association with other *Dryopithecus* fossils since Lartet's day strengthen his conjecture. These cosmopolitan apes evidently preferred wooded tropical and subtropical environments, where they lived by browsing on leaves and fruit.

The first *Dryopithecus* fossils from France consisted of three partial lower jaws, one of which had retained all but one of its teeth. No upper jaws were found, and most other fossils of *Dryopithecus* found in Europe consisted only of isolated teeth. As a result almost nothing was known about the oak ape's skull, face or other body parts until the late 1940's. At that time L. S. B. Leakey and his co-workers in East Africa began to find more complete primate specimens in fossil-rich Miocene deposits on islands in Lake Victoria and at sites inland from the northeastern shore of the lake. These ape remains included parts of jaws that ranged in size from those of living gibbons (the smallest of today's apes) to those of living gorillas. In 1948 Leakey's wife Mary found a beautifully preserved fossil skull on Rusinga Island in Lake Victoria, and three years later the Leakeys' colleague Tom Whitworth discovered parts of a second skull on the island, associated with a forelimb, a hand and some other limb bones, including part of a foot. These fossil African apes were assigned to the genus *Proconsul;* the two skulls and the limb bones were assigned in particular to the species *Proconsul africanus*. Studies in recent years lead to the conclusion that *Proconsul* is not a unique genus but an African member of the cosmopolitan genus *Dryopithecus*.

Two more facts about the dryopithecines should be noted before we follow the branching of man's family tree any further. One is that dryopithecine fossils are found both in Miocene deposits and in late deposits of the preceding epoch, the Oligocene. This means that dryopithecine apes flourished over a period of some 20 million years. The other fact is that such fossils have been found not only in France and East Africa but also at sites in other regions extending across a vast area of the Old World: the western desert of Egypt, the region of Barcelona, the valley of the Rhine, the region of Vienna, the mountains of northeastern Hungary, Macedonian Greece, Asia Minor, the Potwar plateau of Pakistan, the Siwalik Hills of India,

the coalfields of Yunnan in western China and several localities in south-central China.

In the course of their wide and long-lasting radiation these apes seem to have encountered increasingly cooler environments where tropical and subtropical forests gave way to temperate environments with open woodlands and woodland savanna. Fossils found in Europe and Asia since 1970 suggest that between 10 and 15 million years ago *Dryopithecus* gave rise to at least three other genera. Two of them, *Sivapithecus* and *Gigantopithecus*, were primates with a face as large as that of a modern chimpanzee or gorilla. The third genus, *Ramapithecus*, had a small face.

The exact relations among the three advanced Miocene primates are likely to be clarified only when skulls complete with brain cases and other remains such as limb bones are found. *Sivapithecus* has often been classified as a kind of dryopithecine and therefore an ape rather than a hominid. *Ramapithecus* has most often been identified as a member of man's own hominid line. *Sivapithecus* and *Gigantopithecus* do, however, show some hominid characteristics. For example, all three genera have cheek teeth with thick enamel. Such resemblances are presumably due to the fact that the three genera are related, but they may also reflect similar responses to the same environmental changes. Of the three genera, *Ramapithecus* clearly shows the greatest similarity to later hominids. Nevertheless, the discovery of more fossils of *Sivapithecus* and *Gigantopithecus*, not to mention more of *Ramapithecus*, will undoubtedly alter concepts of how manlike animals branched from apelike ones. In any event, let us now turn to some later branches of man's family tree, beginning with *Ramapithecus*.

Haritalyangar is the name given to a cluster of villages in India some 100 miles north of New Delhi in the Siwalik Hills, an area where exposed Miocene fossil beds extend from northwestern India into adjacent Pakistan. There in

1932 G. Edward Lewis, a young Yale University graduate student who had gone into the area alone with a packhorse, discovered fossils of what he called "manlike apes." He assigned one of the fossils, an upper jaw, to a new genus and species he named *Ramapithecus brevirostris*. The generic name simply means "Rama's ape," Rama being the mythical prince who is the hero of an Indian epic poem. The species name that Lewis chose was more meaningful; it is the Latin for "short-snouted," a feature uncharacteristic of apes. Lewis was impressed by many manlike aspects of the jaw and its array of teeth. Writing his doctoral dissertation in 1937, he placed the new genus in the family Hominidae, the division of the order Primates whose sole living representative is *Homo sapiens*.

The next *Ramapithecus* fossil find was not made until 1961. Discovered by L. S. B. Leakey near Fort Ternan in southwestern Kenya, it included parts of both sides of an upper jaw. Leakey, with customary panache, treated his discovery as though it were a totally new genus and species, giving it the name *Kenyapithecus wickeri* after his friend Fred Wick-

er, on whose farm the fossil was found. With hindsight it seems likely that if a lower jaw had been found in direct association with either the Haritalyangar or Fort Ternan upper jaws, or if Leakey had been able to compare his discovery directly with the specimen from Haritalyangar, housed at Yale, there would later have been fewer generic names to deal with.

This, however, is by no means certain. The next *Ramapithecus* specimen to surface after that, which had been excavated in Greece during World War II but was not formally described until 1972, was assigned to another new genus and species: *Graecopithecus freyburgi*. The species name honored Bruno von Freyburg, a German geologist who had pried the fossil out of a Miocene stratum while he was stationed in Athens during the German occupation of Greece. Von Freyburg's find was the complete toothbearing part of a lower jaw, and at the time of its discovery it contained all the teeth. It was considered relatively unimportant because another scholar wrongly identified the fossil as being the jaw of *Mesopithecus*, a common Miocene monkey. Before the specimen had even been photographed most of the teeth were

knocked off and lost during a wartime bombing raid on Berlin.

Next to be added to the growing inventory of *Ramapithecus* fossils was a lower jaw unearthed from a Miocene deposit near Çandir, some 40 miles northeast of Ankara in Turkey. The discovery was made in 1973 and the specimen was described about a year later. At that time it was assigned to another genus and an entirely new species: *Sivapithecus alpani*. *Sivapithecus*, or Siva's ape, was the name Lewis had given to one of his Indian specimens; the species name of the Çandir jaw honors the director of the Turkish Geological Survey.

A major group of *Ramapithecus*-like fossils has also been uncovered in coal deposits of Miocene age in the Rudabanya Mountains of northeastern Hungary. The specimens were found over a period of several years but were not described in detail until 1975. They have been assigned to still another new genus and species: *Rudapithecus hungaricus*.

Of all the recently discovered or recently described *Ramapithecus* fossils only those found in Pakistan over the past 18 months by my Yale colleague David R. Pilbeam and his co-workers have been recognized from the start as

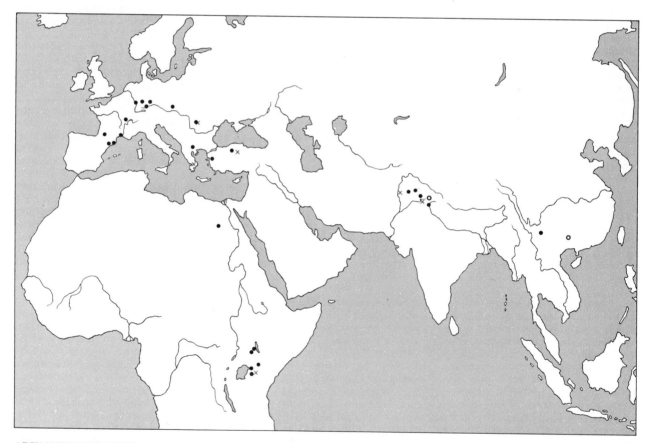

ADVANCED PRIMATES of Oligocene, Miocene and even later times included an abundant and cosmopolitan group of large apes: the dryopithecines (*black dots*). Most may be assigned to two genera: *Dryopithecus*, found mainly in Africa and Europe, and *Sivapithecus*, found mainly from Asia Minor eastward. These great apes flourished for more than 25 million years. The stock gave rise both to the great ape *Gigantopithecus* (*black circles*) and to hominids such as *Ramapithecus*, a genus now known from five areas in Eurasia and one in East Africa (*colored crosses*). Like *Ramapithecus*, *Gigantopithecus* exhibits anatomical changes in the jaw and teeth related to diet.

HIGHER CATEGORIES   GENERA

		LIVING GREAT APES	PONGO (ORANGUTAN) PAN (CHIMPANZEE) GORILLA (GORILLA)
SUPERFAMILY HOMINOIDEA (ALL GREAT AND LESSER APES AND ALL HUMANS AND PREHUMANS)	FAMILY PONGIDAE (ALL GREAT APES)	EXTINCT GREAT APES	DRYOPITHECUS SIVAPITHECUS GIGANTOPITHECUS
	FAMILY HOMINIDAE (ALL HUMANS AND PREHUMANS)	LIVING HUMANS	HOMO (MAN)
		EXTINCT PREHUMANS	RAMAPITHECUS AUSTRALOPITHECUS

SUPERFAMILY HOMINOIDEA, a major subdivision of the order Primates, consists of three families; two of them, the pongids and the hominids, are shown in the diagram. The family Pongidae includes, in addition to three genera of living great apes, the three fossil genera shown. The family Hominidae includes the living genus *Homo* and two extinct prehuman genera.

belonging to *Ramapithecus*. Even a fragment of a lower jaw from Fort Ternan, reclassified in 1971 as *Ramapithecus* by Peter Andrews of the British Museum (Natural History), was originally believed to represent *Dryopithecus*.

Not all this profligate naming was due to overemphasis on what taxonomists call "splitting" (in contrast to "lumping"). Ibrahim Tekkaya, the Turkish paleontologist who assigned the Çandir jaw to *Sivapithecus*, had little material for comparison. The principal information at his disposal was a discussion of a *Dryopithecus* jaw that had been mistakenly assigned to *Ramapithecus* in 1938. This mistaken assignment had been persistently repeated in the scientific literature, and Tekkaya thought the lack of similarity between that jaw and his find disqualified the find as a *Ramapithecus* specimen. He and Andrews have since published a joint report correctly reclassifying the find as *Ramapithecus*.

By the same token Miklos Kretzoi, the Hungarian paleontologist who named his new finds *Rudapithecus*, might not have done so if the first lower jaw retrieved in Hungary in 1969 had not been heavily eroded and incomplete. With the later Rudabanya fossils in hand (these include both upper and lower jaws) the resemblance to *Ramapithecus* is unmistakable. Kretzoi now recognizes his fossils as being "ramapithecine" even if they are not generically *Ramapithecus*.

Finally, the eminent paleoanthropologist G. H. R. von Koenigswald, who named the Athens lower jaw *Graecopithecus*, separated it from *Ramapithecus* because relatively complete lower jaws of *Ramapithecus* were then unknown. The first good specimen to be found, the Çandir mandible, was not described until 1974. Therefore even as recently as 1972 von Koenigswald could not have perceived the close similarities that are evident when the Athens specimen is compared with the ones from Fort Ternan and Çandir.

There are still other "rehabilitated" remains of *Ramapithecus*. For example, Pilbeam and I, writing in 1965, followed the received opinion that hominids do not have the "simian shelf" that is characteristic of apes and monkeys. The simian shelf is a torus, or horizontal ridge of bone, that projects inward from the inside of the lower jaw. Its presence in a half lower jaw from near Domeli in Pakistan, we thought at the time, excluded the fossil from consideration as being a specimen of *Ramapithecus*. We later realized that shelflike ridges are present inside the lower jaw of both *Ramapithecus* and the later African hominid *Australopithecus*. In 1969 Pilbeam rescued the Domeli mandible from its undeserved obscurity, enumerated its hominid features and added it to the inventory of *Ramapithecus* fossil remains.

At this writing some two dozen of these assorted *Ramapithecus* jaws and teeth have been formally described. What do they tell us about man's most remote hominid ancestor? The evidence consists of a number of anatomical fine points, but they lead to an inevitable conclusion: *Ramapithecus* had adapted to a way of life quite different from that followed by most of its forest-dwelling relatives of the *Dryopithecus* group. Where its jaws and teeth reflect that adaptation they resemble those of the African hominid *Australopithecus*. Between the two, however, is a large gap in time. *Ramapithecus* species are not known to have flourished in Eurasia more recently than eight million years ago. The earliest fossils of *Australopithecus* and *Homo* appear to be those found since 1974 in Ethiopia by Donald C. Johanson of Case Western Reserve University and in Tanzania by Mary Leakey; of these the most ancient are all less than four million years old.

To turn to the anatomical details, in 1967, while I was reviewing the jaw morphology of the best-known fossil apes, I found to my surprise that apes in the Miocene epoch had a dental arcade (the arch formed by the row of teeth seen from above) quite unlike that of typical living apes. Among the living apes the arcade is *U*-shaped because the two sides of the jaw are parallel. In typical Miocene apes the arcade is *V*-shaped because the two sides of the jaw diverge to the rear.

Two years later E. Genet-Varcin, writing in Paris, pointed out that the dental arcade of most *Australopithecus* jaws is also *V*-shaped. (In both groups the point of the *V* is squared off in front.) Thus the same kind of arcade is shared by two groups separated by some millions of years. The dental arcade of *Homo sapiens* is neither *U*-shaped nor *V*-shaped but more semicircular in outline. Therefore both the semicircular arcade of modern man and the *U*-shaped one of modern apes presumably arose from a *V*-shaped original.

Where do the jaws of *Ramapithecus* fit into this range of variations? It turns out that they are intermediate. The rows of teeth diverge about 20 degrees from the parallel. This angle is greater than that in Miocene apes (about 10 degrees) and less than that in *Australopithecus* (about 30 degrees).

Both *Ramapithecus* and *Australopithecus* have a lower jaw that is typically about as thick, from the cheek side to the tongue side, in the molar region as it is deep from top to bottom. This robustness is in contrast to the lower jaw of living apes, which in the molar region is much thinner with respect to its depth. The molar teeth of *Ramapithecus* and *Australopithecus* convey the same impression of robustness. They are large and flat, and viewed from above most of them have rounded outlines. Unlike the molars of chimpanzees and gorillas, and of *Dryopithecus* in the narrow sense (excluding *Sivapithecus*), they have a thick enamel and hence are more resistant to wear. The canine teeth and incisors are smaller than those of living apes. Furthermore, the lower jaw of *Ramapithecus* and *Australopithecus* is buttressed against the stresses of heavy chewing by not one shelf on the inside but two. The shelves, known respectively as the superior and the inferior torus, were still present in the more robust species of *Australopithecus* as recently as a million years ago.

Another unusual feature shared by these two hominids is a marked difference in the amount of wear shown by the first and second molars. In *Dryopithecus* and modern apes when the enamel on the cusps of the first molar is worn away to expose the underlying dentine, the cusps of the second molar usually show nearly as much erosion. In *Ramapithecus* and *Australopithecus* the difference in wear between the first and second molars is evidence that the eruption of these permanent teeth was separated by a considerable length of time. This in turn

a

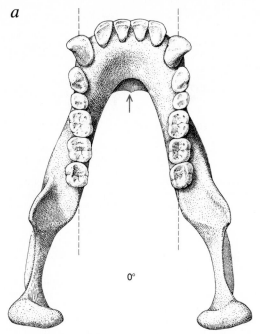

0°

b

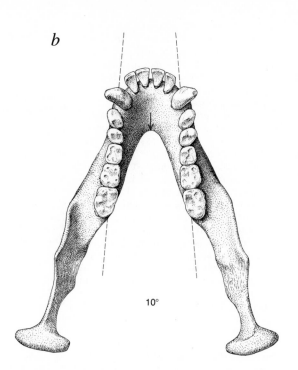

10°

FOUR LOWER JAWS show variations in the amount of rearward divergence of the tooth arcades in three fossil primates. At far left (*a*), for comparison, is the mandible of a modern chimpanzee; its typically *U*-shaped dental arcade has parallel tooth rows; thus the degree of divergence is zero. Next (*b*) is a reconstructed *Dryopithecus* mandible; the tooth rows show an angle of divergence (*color*) averaging some 10 degrees. Next (*c*) is a composite reconstruction of a *Ramapithecus* mandible. Its tooth rows, when preserved, show an angle of

may mean that both hominids matured slower than Miocene or modern apes.

This group of related features—thickened and buttressed jaws, flattened molars and reduced canines and incisors—can be compared with the dental system of certain other mammals: hoofed herbivores, herbivorous rodents and even the elephant. These mammals chew mostly by means of a rocking or grinding action, so that the movement of the teeth is partly sideways. Where chewing by chopping and biting mainly exerts vertical forces, grinding includes some nearly horizontal forces. A further adaptation in these mammals is a vertical lengthening of the ascending ramus, or upward extension, at the rear of the lower jaw, which is set at a right angle. This contrasts sharply with the short and backward-sloping ascending ramus of modern man and the chimpanzee. Man and the chimpanzee also lack thick tooth enamel and robust jaws.

It is therefore not surprising to find that the ascending ramus of the lower jaw in both *Ramapithecus* and *Australopithecus* is typically at more of a right angle to the horizontal than that in modern man and the living apes of Africa. In *Ramapithecus* and *Australopithecus* the ascending ramus also seems to have been proportionately higher. The front of the ramus, the coronoid process, is shifted slightly forward with respect to the cheek teeth. This angularity and shifting is correlated with changes in the upper jaw and lower face. The bone of

the upper jaw is deepened and thickened around the roots of the teeth, and the snout is flat and short, with the nasal openings located close to the upper incisors. The cheekbones are set farther forward, and the upper incisors and canine teeth tend to be directed downward rather than forward. Such striking similarities between two genera so widely separated in time must have some meaning.

The first intimations of a shift in a major adaptation of an organism, reflected in a change in the function of an existing anatomical system, provide some of the clearest signposts in the fossil record. The anatomical change often signals a dramatic evolutionary event to come. In primate evolution one of the latest of these adaptive shifts and perhaps the most significant is the one that gave rise to the quantum advance that sets man apart from the other primates. I believe this key shift is the one intimated by the functional change in the dental system of *Ramapithecus*.

When and why did *Ramapithecus* diverge from its cosmopolitan relatives, the dryopithecines, developing a jaw anatomy foreshadowing that of the two later hominid genera, *Australopithecus* and *Homo?* We can begin by looking for places where *Dryopithecus* encountered altered environmental circumstances that would have increased the probability of such a divergence.

Among the dryopithecines the group

with the closest ties to *Ramapithecus* seems to be *Sivapithecus,* which is mainly Asian in distribution; the African dryopithecines were considerably earlier. Many of the more recent fossil apes from Eurasia are rather large, with massive, deep jaws and thickened tooth enamel; such specimens can be placed in the genus *Sivapithecus.* Two of these Eurasian fossil apes are now well known. One of them, *Sivapithecus indicus,* is represented by remains found mainly in northern India, Kashmir, Pakistan and Turkey. The other, which probably belongs to a new and still undescribed genus, is represented by a find from northern Greece that was named "*Dryopithecus" macedoniensis* in 1974. Its discovery was followed by the unearthing at the same location of several more partial jaws of the same species; all resemble *Sivapithecus* more than they do *Dryopithecus.* Some of the individuals among these far-flung populations of *Sivapithecus* and its close relatives have a face comparable in size to that of a gorilla; very few have a face as small as that of a chimpanzee.

In late Miocene times, 10 to 12 million years ago, much of Eurasia was covered with forest. The cover was not, however, tropical forest. Thus it did not provide the kind of year-long fruit production and continuous vegetative renewal that is typical of the relatively seasonless forests where apes now reside. Under those circumstances it seems likely that the larger of the Mio-

c   *d*

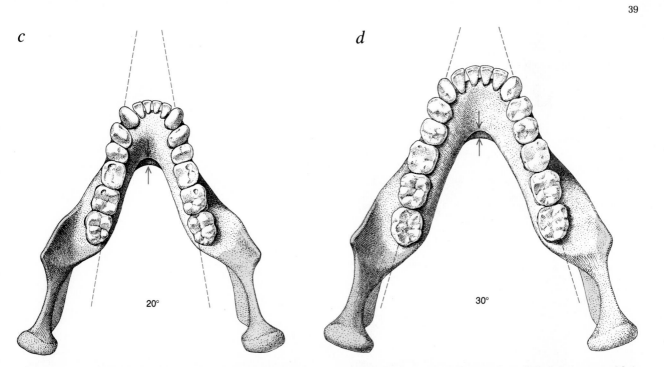

20°   30°

divergence averaging 20 degrees. Last (*d*) is a reconstructed *Australopithecus* mandible. Its typical angle of tooth-row divergence is 30 degrees. The tooth rows of later hominids show even greater angles of divergence. Colored arrows show differences in the two jaw-ridge buttresses known as the superior and the inferior torus. Modern apes possess a large, shelflike inferior torus; in *Dryopithecus* the superior torus was dominant. Both of the ridges are developed in *Ramapithecus* and *Australopithecus* (*see illustration on next page*).

cene apes would often have found that the food available to them in the trees was inadequate to their needs. A tendency to pursue an alternative food-gathering strategy—foraging on the ground and along the edge of the forest for small, tough foods such as nuts or roots—could have provided natural-selection pressures favoring the survival of individuals with more robust jaws and thickened tooth enamel.

The animal remains associated with *Ramapithecus* at certain sites in Eurasia suggest an even more distinct forest-fringe and wooded-savanna adaptation. For example, the animal remains at Çandir include species that were at home in both forest and savanna environments. Although *Ramapithecus* was present in those habitats, no Miocene apes were. The animals in the Athens Miocene formation are species indicative of a savanna-and-steppe environment; there too *Ramapithecus* was present but apes were not. In Hungary, in Pakistan and in India the fossil plants and animals associated with *Ramapithecus* suggest subtropical to warm-temperate environments, and there *Ramapithecus* and apes lived together. Such habitats could have supported mixed populations of higher primates, including not only *Ramapithecus* but also the larger *Sivapithecus* and *Gigantopithecus*, ranging in size from the size of a chimpanzee to that of a gorilla. These Miocene primates would have been in the midst of the dietary shift that is recorded in their

altered dental system. Even as they were still living in forests and on the margins of forests they had given up feeding in the trees in favor of feeding on the ground. Indeed, *Ramapithecus* was already at home away from the treetops in the open woodland and savanna.

It seems unnecessary to postulate two separate adaptations to the same circumstances, one affecting *Ramapithecus* and the other the larger dryopithecines of Eurasia, in order to account for these anatomical developments. Rather, one can speculate that somewhere in southern Eurasia about 15 million years ago a stock of ground-adapted apes that already showed a tendency to forage on small and tough foodstuffs initiated a divergence that led to the appearance of two groups, one manlike and the other still apelike, with somewhat similar dental systems. Although both groups are now extinct, *Ramapithecus*, the hominid, may well have given rise to the later hominid *Australopithecus*.

As we have seen, a gap of at least four million years, a length of time greater than the documented span of the genus *Homo*, separates the youngest-known fossil remains of *Ramapithecus* from the oldest-known representatives of *Australopithecus*. Although fossils of *Australopithecus* have so far been found only in Africa, could *Australopithecus* have been as cosmopolitan as *Ramapithecus?* This is certainly possible, but the possibility will remain a conjecture until some fossil evidence is unearthed in sup-

port of it. A comparable situation may nonetheless be worth mentioning. The first fossils of *Homo erectus*, the immediate precursor of modern man, were recognized by Eugène Dubois in Java in the 1890's. Thereafter it was not until the early 1930's, when fossils of the same general type were unearthed in China, that the existence of *Homo erectus* outside Java was demonstrated. Many more years were to pass before fossil forms of *Homo erectus* were recognized in southern, eastern and northern Africa and possibly in Europe as well. Today the genus is seen as having a true cosmopolitan status. Thus it would not be surprising if fossils of *Australopithecus* were found on other continents. What would be surprising, biologically speaking, would be if only that part of the cosmopolitan *Ramapithecus* population living in Africa gave rise to *Australopithecus*.

So far this tracing of pathways has led from generalized hominoids to the hominids by way of an unknown offshoot of *Dryopithecus* that gave rise both to the large ground apes of southern Eurasia and to the hominid *Ramapithecus*. This hominid in turn may well have given rise to the hominid genus *Australopithecus*. What pathway leads to the genus *Homo?*

Here again we find resemblances between *Ramapithecus* and the most primitive recognized species of *Homo*, although the resemblances are not always the same. The name *Homo habilis*,

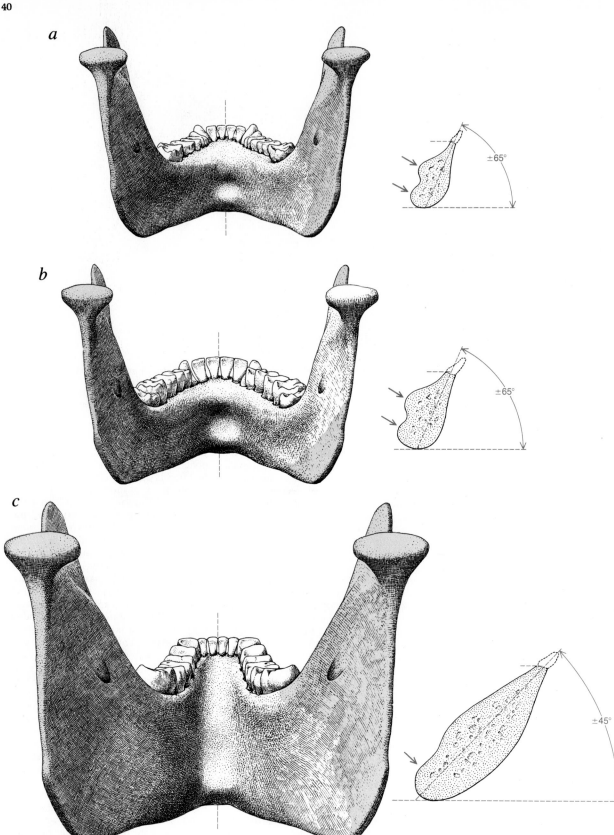

THREE FOSSIL MANDIBLES are seen from the rear to show the location of the bony jaw buttresses indicated in the cross sections (*right*). The sections show the plane where the left and right halves of the mandibles meet. The buttresses evidently strengthen the mandible against the lateral stresses produced by side-to-side chewing. In *Ramapithecus* (*a*) and *Australopithecus* (*b*) the buttressing is much the same. *Gigantopithecus* (*c*) more resembles the Miocene ape *Siva-* *pithecus* and the modern gorilla: the lower buttress is more developed than the upper. In the two hominids the long axis of the cross section is also at a relatively high angle with respect to the horizontal; in the specimens shown here it is about 65 degrees, but it is often higher. This contrasts sharply with the *Gigantopithecus* cross section; the entire front end of that mandible is lengthened and tilted forward, making a lower angle with the horizontal (45 degrees in this example).

a

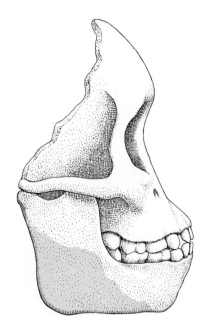

b

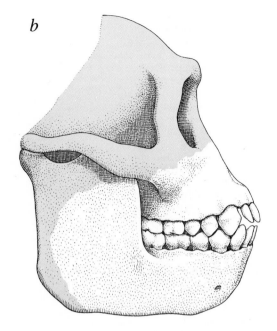

c

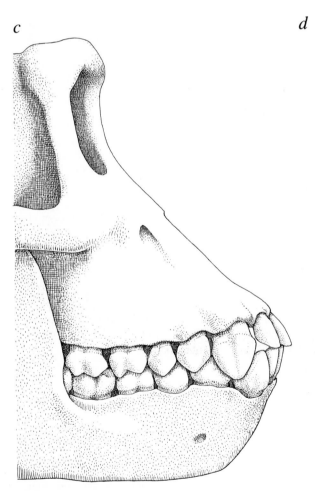

d

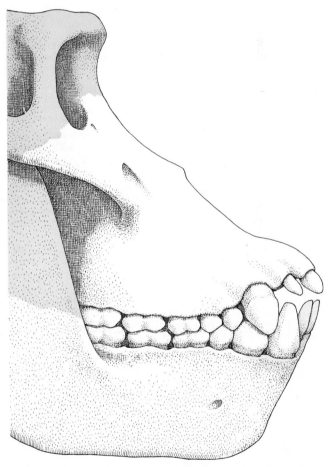

PROFILES OF FOUR PRIMATES contrast the size and structure of their faces. First is the Taung juvenile (*a*), the specimen of *Australopithecus* first described by Raymond Dart in 1925. Second is a reconstruction of *Ramapithecus* (*b*), a hominid hardly larger than the *Australopithecus* juvenile. The profile is based on recently discovered specimens (*color indicates conjectural areas*). Third is the face of a modern female gorilla (*c*) and fourth is a reconstruction of a large contemporary of *Ramapithecus*, the Miocene ape *Sivapithecus* (*d*). Changes in the anatomy of the upper and lower jaws and the teeth of *Ramapithecus*, which evidently relate to diet, have made the hominid less snouty than either fossil or living apes and more like *Australopithecus* juveniles and adults (*see illustration on pages 38 and 39*).

coined in 1964 by Leakey to categorize certain fossils from his famous site at Olduvai Gorge in Tanzania, is usually applied to a group of African fossils that many students of human evolution believe represent the forerunners of *Homo erectus*. They may also include the prehuman fossils recently found by Mary Leakey in Tanzania and some of the fossils uncovered by Johanson in Ethiopia; both groups of specimens are believed to be between three and four million years old. The specimens from Tanzania and Ethiopia resemble somewhat younger fossils of primitive members of the genus *Homo* uncovered in northeastern Kenya by the Leakeys' son Richard E. Leakey, including the well-known cranium and face designated E.R. (for East Rudolf) 1470.

Homo habilis is a name that needs to be overhauled. Few workers agree on what specimens belong in the category. Nevertheless, it is clear that many of these early members of the genus *Homo* in Africa had canine and incisor teeth somewhat larger than those typical of the specimens of *Australopithecus* that are contemporary with them. There are other apparently correlated differences between the early members of the genus *Homo* and *Australopithecus:* the brain of the *Homo* specimens is somewhat larger, and so is their body. The emergence of *Homo* from *Australopithecus,* however, is a complex subject that will not be taken up here. What remains to be dealt with are the resemblances between *Ramapithecus* and primitive forms of *Homo,* such as *Homo habilis* and *Homo erectus,* that are not seen in typical specimens of *Australopithecus.*

Here again the evidence mainly has to do with jaws and teeth. That evidence was first pointed out by von Koenigswald in relation to a *Homo erectus* specimen found by his collectors at the Sangiran site in Java in 1939. The specimen is usually designated either Sangiran II or Pithecanthropus IV. As in the earliest African species of *Homo* from Ethiopia, Kenya and Tanzania (and in L. S. B. Leakey's first specimen of *Homo habilis*), the upper canine teeth of Sangiran II are comparatively large. Moreover, whereas many *Australopithecus* premolars are "molarized," that is, enlarged almost to the size of molars, the premolars of Sangiran II are not. The upper canine tooth of Sangiran II overlapped the teeth of the lower jaw to such an extent that it caused facets of wear to appear on the opposing canine and first premolar in the lower jaw. There is also a diastema, or small gap, in front of the upper canine, separating it from the next tooth forward, the lateral incisor.

Such gaps between the teeth have traditionally been viewed as an apelike characteristic. For many years the San-

FOSSIL SITE IN INDIA, near the area where G. Edward Lewis first discovered *Ramapithecus* in 1932, was visited by a group of paleontologists from Yale University in 1969. The site, a part of the fossil-rich Miocene formation in the Siwalik Hills, contained *Ramapithecus* teeth.

giran II diastema was often explained away as an anomaly. One of the best of the new finds from Ethiopia, a complete palate with upper teeth, shows that these dental relations are not anomalous: the fit of the upper teeth closely resembles the fit of the same teeth in *Ramapithecus*. In contrast, the canines of a typical *Australopithecus* specimen are so small that they do not overlap and cause wear, and there is almost no diastema in the upper jaw. It should also be pointed out that where the upper-jaw diastema is found in *Homo erectus* and *Homo habilis* (and in *Ramapithecus*) the teeth of the lower jaw are closely packed together. This close packing is typical of *Australopithecus* but, as we have seen, not of apes.

A further observation can be made. The earliest fossils of *Homo* from Africa such as E.R. 1470, if indeed they should be designated *Homo,* show a facial and cranial modeling that is quite reminiscent of *Australopithecus*. If it were not for such emergent human characteristics as an enlarged brain and teeth resembling those of *Homo erectus,* all these fossils might well be classified as another species of *Australopithecus* that could be called *Australopithecus habilis*.

The simplest way to resolve these complex relations is to postulate that the hominid stock ancestral both to primitive *Homo* and to *Australopithecus* resem-

bled *Ramapithecus* more closely than later representatives of *Australopithecus* did. The resemblance centers on the possession of large canine teeth. The only evolutionary room available in the fossil record for such a postulated ancestral form is the period between the last appearance of *Ramapithecus* and the first appearance of *Homo* and *Australopithecus*. This is the period between four and eight million years ago, or exactly where there is now a large gap in the fossil record.

There is an alternative to this simple resolution of the matter, and it is one that L. S. B. Leakey favored. It postulates that *Homo* and *Australopithecus* branched independently and directly from *Ramapithecus*. There is a third and less likely possibility that does not fit with the interpretation of *Ramapithecus* presented here. It is that not only *Ramapithecus* but also other hominids emerged independently from the parental stock of Miocene apes. If future discoveries should prove this to be the case, and significant gaps in present knowledge prevent its being entirely ruled out, then current conceptions of the origin of the hominids will have to be drastically revised. Even if that should happen, however, *Ramapithecus* would almost certainly retain its position as a very early hominid.

The Evolution of the Hand

4

by John Napier
December 1962

In 1960 tools were found together with the hand bones
of a prehuman primate that lived more than a million
years ago. This indicates that the hand of modern
man has much later origins than had been thought

At Olduvai Gorge in Tanganyika two years ago L. S. B. Leakey and his wife Mary unearthed 15 bones from the hand of an early hominid. They found the bones on a well-defined living floor a few feet below the site at which in the summer of 1959 they had excavated the skull of a million-year-old man-ape to which they gave the name *Zinjanthropus*. The discovery of *Zinjanthropus* has necessitated a complete revision of previous views about the cultural and biological evolution of man. The skull was found in association with stone tools and waste flakes indicating that at this ancient horizon toolmakers were already in existence. The floor on which the hand bones were discovered has also yielded stone tools and a genuine bone "lissoir," or leather working

tool. Hence this even older living site carries the origins of toolmaking still further back, both in time and evolution, and it is now possible for the first time to reconstruct the hand of the earliest toolmakers.

Research and speculation on the course of human evolution have hitherto paid scant attention to the part played by the hand. Only last year I wrote: "It is a matter of considerable surprise to many to learn that the human hand, which can achieve so much in the field of creative art, communicate such subtle shades of meaning, and upon which the pre-eminence of *Homo sapiens* in the world of animals so largely depends, should constitute, in a structural sense, one of the most primitive and generalized parts of the human body." The im-

plication of this statement, which expresses an almost traditional view, is that the primate forebears of man were equipped with a hand of essentially human form long before the cerebral capacity necessary to exploit its potential had appeared. The corollary to this view is that the difference between the human hand and the monkey hand, as the late Frederic Wood Jones of the Royal College of Surgeons used to insist, is largely one of function rather than structure. Although broadly speaking it is true that the human hand has an extraordinarily generalized structure, the discovery of the Olduvai hand indicates that in a number of minor but nevertheless highly significant features the hand is more specialized than we had supposed.

Tool-using—in the sense of improvisa-

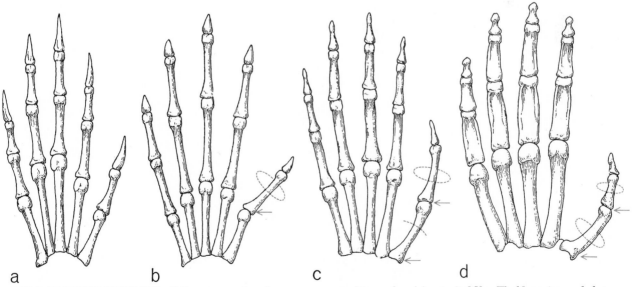

HANDS OF LIVING PRIMATES, all drawn same size, show evolutionary changes in structure related to increasing manual dexterity. Tree shrew (*a*) shows beginnings of unique primate possession, specialized thumb (*digit at right*). In tarsier (*b*) thumb is distinct and can rotate around joint between digit and palm. In capuchin monkey (*c*), a typical New World species, angle between thumb and finger is wider and movement can be initiated at joint at base of palm. Gorilla (*d*), like other Old World species, has saddle joint at base of palm. This allows full rotation of thumb, which is set at a wide angle. Only palm and hand bones are shown here.

tion with naturally occurring objects such as sticks and stones—by the higher apes has often been observed both in the laboratory and in the wild and has even been reported in monkeys. The making of tools, on the other hand, has been regarded as the major breakthrough in human evolution, a sort of status symbol that could be employed to distinguish the genus *Homo* from the rest of the primates. Prior to the discovery of *Zinjanthropus*, the South African man-apes (Australopithecines) had been associated at least indirectly with fabri-

cated tools. Observers were reluctant to credit the man-apes with being tool-makers, however, on the ground that they lacked an adequate cranial capacity. Now that hands as well as skulls have been found at the same site with undoubted tools, one can begin to correlate the evolution of the hand with the stage of culture and the size of the brain. By the same token one must also consider whether the transition from tool-using to toolmaking and the subsequent improvement in toolmaking techniques can be explained purely in

terms of cerebral expansion and the refinement of peripheral neuromuscular mechanisms, or whether a peripheral factor—the changing form of the hand—has played an equally important part in the evolution of the human species. And to understand the significance of the specializations of the human hand, it must be compared in action—as well as in dissection—with the hands of lower primates.

In the hand at rest—with the fingers slightly curled, the thumb lying in the plane of the index finger, the poise of the

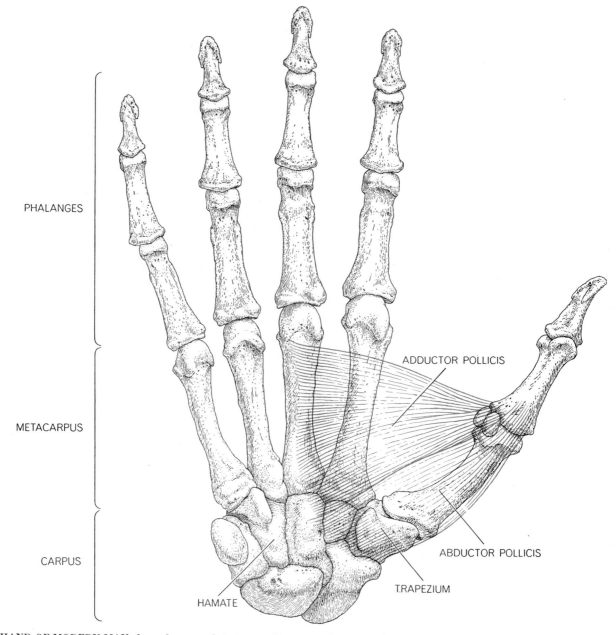

PHALANGES

METACARPUS

CARPUS

ADDUCTOR POLLICIS

ABDUCTOR POLLICIS

T. RAPEZIUM

HAMATE

HAND OF MODERN MAN, drawn here actual size, is capable of precise movements available to no other species. Breadth of terminal phalanges (end bones of digits) guarantees secure thumb-to-finger grip. Thumb is long in proportion to index finger and is set at very wide angle. Strong muscles (*adductor pollicis* and *abductor pollicis*) implement movement of thumb toward and away from palm. Saddle joint at articulation of thumb metacarpal (a bone of the palm) and trapezium (a bone of the carpus, or wrist) enables thumb to rotate through 45 degrees around its own longitudinal axis and so be placed in opposition to all the other digits.

whole reflecting the balanced tension of opposing groups of muscles—one can see something of its potential capacity. From the position of rest, with a minimum of physical effort, the hand can assume either of its two prehensile working postures. The two postures are demonstrated in sequence by the employment of a screw driver to remove a screw solidly embedded in a block of wood [*see illustration below*]. The hand first grips the tool between the flexed fingers and the palm with the thumb reinforcing the pressure of the fingers; this is the "power grip." As the screw comes loose, the hand grasps the tool between one or more fingers and the thumb, with the pulps, or inner surfaces, of the finger and thumb tips fully opposed to one another; this is the "precision grip." Invariably it is the nature of the task to be performed, and not the shape of the tool or object grasped, that dictates which posture is employed. The power grip is the grip of choice when the full strength of the hand must be applied and the need for precision is subordinate; the precision grip comes into play when the need for power is secondary to the demand for fine control.

The significance of this analysis becomes apparent when the two activities are correlated with anatomical structure. The presence or absence of these structural features in the hands of a lower primate or early hominid can then be

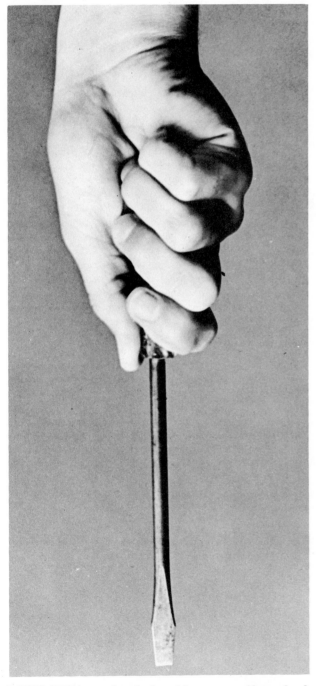

POWER GRIP is one of two basic working postures of human hand. Used when strength is needed, it involves holding object between flexed fingers and palm while the thumb applies counterpressure.

PRECISION GRIP is second basic working posture and is used when accuracy and delicacy of touch are required. Object is held between tips of one or more fingers and the fully opposed thumb.

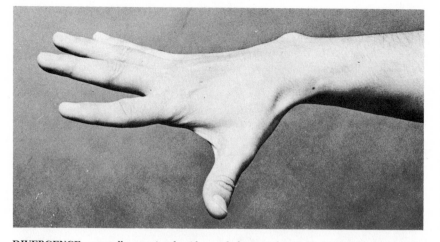

DIVERGENCE, generally associated with weight-bearing function of hand, is achieved by extension at the metacarpophalangeal joints. All mammalian paws are capable of this action.

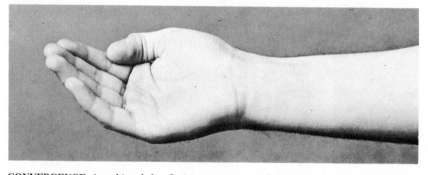

CONVERGENCE is achieved by flexion at metacarpophalangeal joints. Two convergent paws equal one prehensile hand; many mammals hold food in two convergent paws to eat.

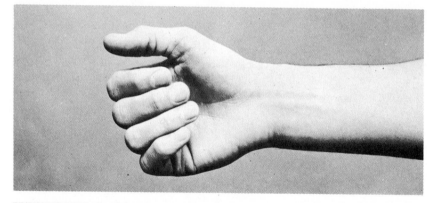

PREHENSILITY, the ability to wrap the fingers around an object, is a special primate characteristic, related to the emergence of the specialized thumb during evolutionary process.

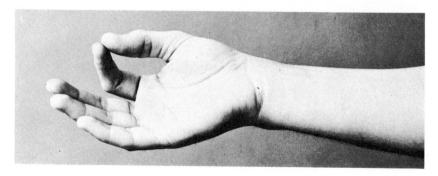

OPPOSABILITY is ability to sweep thumb across palm while rotating it around its longitudinal axis. Many primates can do this, but underlying structures are best developed in man.

taken to indicate, within limits, the capabilities of those hands in the cultural realm of tool-using and toolmaking. In the case of the hand, at least, evolution has been incremental. Although the precision grip represents the ultimate refinement in prehensility, this does not mean that more primitive capacities have been lost. The human hand remains capable of the postures and movements of the primate foot-hand and even of the paw of the fully quadrupedal mammal, and it retains many of the anatomical structures that go with them. From one stage in evolution to the next the later capability is added to the earlier.

The study of primate evolution is facilitated by the fact that the primates now living constitute a graded series representative of some of its principal chapters. It is possible, at least, to accept a study series composed of tree shrews, tarsiers, New World monkeys, Old World monkeys and man as conforming to the evolutionary sequence. In comparing the hands of these animals with one another and with man's, considerable care must be taken to recognize specializations of structure that do not form part of the sequence. Thus the extremely specialized form of the hand in the anthropoid apes can in no way be regarded as a stage in the sequence from tree shrew to man. The same objection does not apply, however, to certain fossil apes. The hand of the Miocene ancestral ape *Proconsul africanus* does not, for example, show the hand specializations of living apes and can legitimately be brought into the morphological sequence that branches off on the man-ape line toward man.

In the lowliest of the living primates —the tree shrew that inhabits the rain forests of the East Indies and the Malay Archipelago—the hand is little more than a paw. It exhibits in a primate sense only the most rudimentary manual capability. This is the movement of convergence that brings the tips of the digits together by a flexion of the paw at the metacarpophalangeal joints, which correspond in man to the knuckles at the juncture of the fingers and the rest of the hand. The opposite movement— divergence—fans the digits outward and is related to the pedal, or weight-bearing, function of the paw. With its paws thus limited the tree shrew is compelled to grasp objects, for example its insect prey, in two-handed fashion, two convergent paws being the functional equivalent of a prehensile hand. For purposes of locomotion in its arboreal

STONE TOOLS to left of center are similar to those found at Olduvai Gorge, Tanganyika, in conjunction with the hand bones of an early hominid. Such crude tools can be made by using the power grip, of which the Olduvai hand was capable. Finely flaked Old Stone Age tools at right can be made only by using the precision grip, which may not have been well developed in Olduvai hand.

habitat, this animal does not require prehensility because, like the squirrel, it is small, it has claws on the tips of its digits and is a tree runner rather than a climber. Even in the tree shrew, however, the specialized thumb of the primate family has begun to take form in the specialized anatomy of this digit and its musculature. Occasionally tree shrews have been observed feeding with one hand.

The hand of the tarsier, another denizen of the rain forests of the East Indies, exhibits a more advanced degree of prehensility in being able to grasp objects by bending the digits toward the palm. The thumb digit also exhibits a degree of opposability to the other digits. This is a pseudo opposability in that the movement is restricted entirely to the metacarpophalangeal joint and is therefore distinct from the true opposability of man's thumb. The movement is facilitated by the well-developed abductor and adductor muscles that persist in the hands of the higher primates. With this equipment the tarsier is able to support its body weight on vertical stems and to grasp small objects with one hand.

The tropical rain forests in which these animals live today are probably not very different from the closed-canopy forests of the Paleocene epoch of some 70 million years ago, during which the first primates appeared. In the wide variety of habitats that these forests provide, ecologists distinguish five major strata, superimposed like a block of apartments. From the top down these are the upper, middle and lower stories (the last being the main closed canopy), the shrub layer and the herb layer on the ground. To these can be added a sixth deck: the subterrain. In the emergence of prehensility in the primate line the three-dimensional arrangement of this system of habitats played a profound role. Prehensility is an adaptation to arboreal life and is related to climbing. In animals that are of small size with respect to the branches on which they live and travel, such as the tree shrew, mobility is not hampered by lack of prehensility. They can live at any level in the forest, from the forest floor to the tops of the tallest trees, their stability assured by the grip of sharp claws and the elaboration of visual and cerebellar mechanisms.

The tree-climbing as opposed to the tree-running phase of primate evolution may not have begun until the middle of the Eocene, perhaps 55 million years ago. What environmental pressure brought about this adaptation can only be guessed at. Thomas F. Barth of the University of Chicago has suggested that the advent of the widely successful order of rodents in the early Eocene may have led to the displacement of the primates from the shrub strata to the upper three strata of the forest canopy. In any case little is known about the form of the primates that made this transition.

In *Proconsul,* of the early to middle Miocene of 20 million years ago, the fossil record discloses a fully developed tree-climbing primate. His hand was clearly prehensile. His thumb, however, was imperfectly opposable. Functionally this hand is comparable to that of some of the living New World monkeys.

True opposability appears for the first time among the living primates in the Old World monkeys. In these animals the carpometacarpal joint shows a well-developed saddle configuration comparable to that in the corresponding joint of the human hand. This allows rotation of the thumb from its wrist articulation. Turning about its longitudinal axis through an angle of about 45 degrees, the thumb can be swept across the palm, and the pulp of the thumb can be directly opposed to the pulp surfaces of one of or all the other digits. This movement is not so expertly performed by the monkeys as by man. At the same time, again as in man, a fair range of movement is retained at the metacarpophalangeal joint, the site of pseudo opposability in the tarsier.

The hands of anthropoid apes display many of these anatomical structures but do not have the same degree of functional capability. This is because of certain specializations that arise from the fact that these apes swing from trees by their hands. Such specializations would seem to exclude the apes from the evolutionary sequence that leads to man. In comparing the hand of monkeys with the hand of man one must bear in mind an obvious fact that is all too often overlooked: monkeys are largely quadrupedal, whereas man is fully bipedal. Variations in the form of the hand from one species of monkey to the next are related to differences in their mode of locomotion. The typical monkey hand is rather long and narrow; the metacarpal, or "palm," bones are short compared with the digits (except in baboons); the terminal phalanges, or finger-tip bones, are slender and the tips of the fingers are consequently narrow from side to side. These are only the most obvious differences between the foot-hand of the Old World monkey and that of man. They serve nonetheless to show how too rigid an application of Frederic Wood Jones's criterion of morphological similarity can mislead one into assuming that the only important difference between the hands of men and monkeys lies in the elaboration of the central nervous system.

It seems likely that the terrestrial phase of human evolution followed on the heels of *Proconsul*. At that time, it is well known, the world's grasslands expanded enormously at the expense of the forests. By the end of the Miocene, 15 million years ago, most of the prototypes of the modern plains-living forms had appeared. During this period, apparently, the hominids also deserted their original forest habitats to take up life on the savanna, where the horizons were figuratively limitless. Bipedal locomotion, a process initiated by life in the trees and the ultimate mechanism for emancipation of the hands, rapidly followed the adoption of terrestrial life. The use of the hands for carrying infants, food and even weapons and tools could not have lagged far behind. As Sherwood L. Washburn of the University of California has suggested on the basis of observations of living higher primates,

tool-using must have appeared at an early stage in hominid evolution. It is a very short step from tool-using to tool-modifying, in the sense of stripping twigs and leaves from a branch in order to improve its effectiveness as a tool or weapon. It is an equally short further step to toolmak-

ing, which at its most primitive is simply the application of the principle of modification to a stick, a stone or a bone. Animal bones are a convenient source of tools; Raymond A. Dart of the University of Witwatersrand in South Africa has advanced the hypothesis that such tools

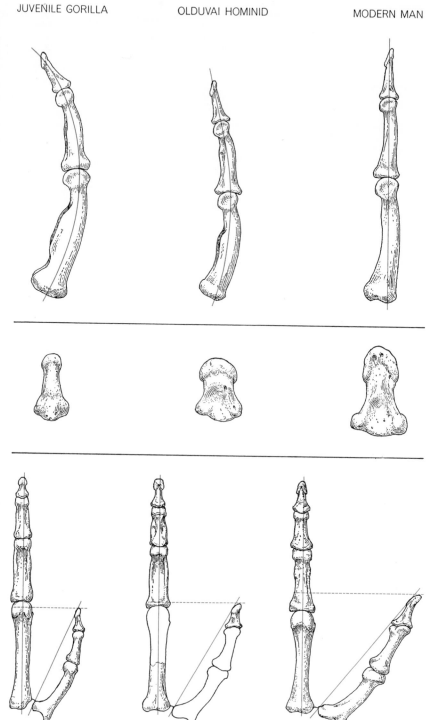

JUVENILE GORILLA   OLDUVAI HOMINID   MODERN MAN

HAND BONES of juvenile gorilla, Olduvai hominid and modern man are compared. Phalanges (*top row*) decrease in curvature from juvenile gorilla to modern man. Terminal thumb phalanx (*middle row*) increases in breadth and proportional length. Third row shows increase in length of thumb and angle between thumb and index finger. Olduvai bones in outline in third row are reconstructed from other evidence; they were not found.

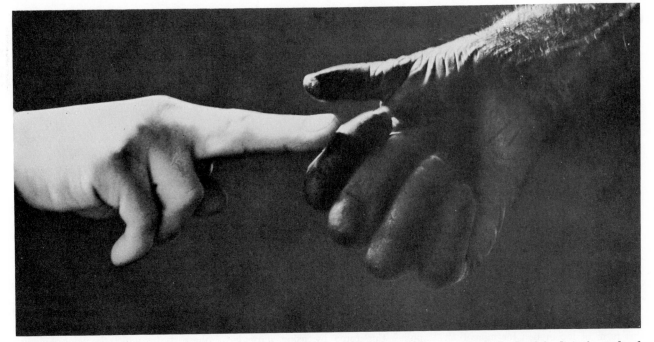

CHIMPANZEE, attempting to grasp experimenter's finger, uses an inefficient precision grip. Because animal's thumb is so short in pro- portion to the digits, it is compelled to bend the digits forward and grasp the object between the sides of index finger and thumb.

were used by early man-apes as part of an "osteodontokeratic" (bone-tooth-hair) culture.

The tools from the pre-*Zinjanthropus* stratum at Olduvai Gorge are little more than pebbles modified in the simplest way by striking off one or more flakes to produce a chopping edge. This technology could not have required either a particularly large brain or a hand of modern human proportions. The hand bones of the pre-*Zinjanthropus* individuals uncovered by the Leakeys in their more recent excavation of Olduvai Gorge are quite unlike those of modern *Homo sapiens*. But there seems to be no reason, on either geological or anthropological grounds, for doubting that the tools found with them are coeval. Modern man must recover from his surprise at the discovery that hands other than his own were capable of shaping tools.

At this point it may be useful to return to the analysis of the manual capability of modern man that distinguishes the power and the precision grip. When compared with the hand of modern man, the Olduvai hand appears to have been capable of a tremendously strong power grip. Although it was a smaller hand, the relative lengths of the metacarpals and phalanges indicate that the proportion of digits and palm was much the same as it is in man. In addition, the tips of the terminal bones of all the Olduvai fingers are quite wide and the finger tips themselves must therefore have been

broad—an essential feature of the human grip for both mechanical and neurological reasons. The curvature of the metacarpals and phalanges indicates that the fingers were somewhat curved throughout their length and were normally held in semiflexion. Unfortunately no hamate bone was found among the Olduvai remains. This wristbone, which articulates with the fifth metacarpal, meets at a saddle joint in modern man and lends great stability to his power grip.

It seems unlikely that the Olduvai hand was capable of the precision grip in its fullest expression. No thumb metacarpal was found in the Olduvai deposit; hence any inference as to the length of the thumb in relation to the other fingers must be derived from the evidence of the position of the wristbone with which the thumb articulates. This evidence suggests that the Olduvai thumb, like the thumb of the gorilla, was set at a narrower angle and was somewhat shorter than the thumb of modern man, reaching only a little beyond the metacarpophalangeal joint of the index finger. Thus, although the thumb was opposable, it can be deduced that the Olduvai hand could not perform actions as precise as those that can be undertaken by the hand of modern man.

Nonetheless, the Olduvai hand activated by a brain and a neuromuscular mechanism of commensurate development would have had little difficulty in making the tools that were found with it. I myself have made such pebble tools

employing only the power grip to hold and strike two stones together.

The inception of toolmaking has hitherto been regarded as the milestone that marked the emergence of the genus *Homo*. It has been assumed that this development was a sudden event, happening as it were almost overnight, and that its appearance was coincidental with the structural evolution of a hominid of essentially modern human form and proportions. It is now becoming clear that this important cultural phase in evolution had its inception at a much earlier stage in the biological evolution of man, that it existed for a much longer period of time and that it was set in motion by a much less advanced hominid and a much less specialized hand than has previously been believed.

For full understanding of the subsequent improvement in toolmaking over the next few hundred thousand years of the Paleolithic, it is necessary to document the transformation of the hand as well as of the brain. Attention can now also be directed toward evidence of the functional capabilities of the hands of early man that is provided by the tools they made. These studies may help to account for the radical changes in technique and direction that characterize the evolution of stone implements during the middle and late Pleistocene epoch. The present evidence suggests that the stone implements of early man were as good (or as bad) as the hands that made them.

5

The Antiquity
of Human Walking

by John Napier
April 1967

*Man's unique striding gait may be the most significant
ability that sets him apart from his ancestors. A big-toe
bone found in Tanzania is evidence that his ability dates
back more than a million years*

Human walking is a unique activity during which the body, step by step, teeters on the edge of catastrophe. The fact that man has used this form of locomotion for more than a million years has only recently been demonstrated by fossil evidence. The antiquity of this human trait is particularly noteworthy because walking with a striding gait is probably the most significant of the many evolved capacities that separate men from more primitive hominids. The fossil evidence—the terminal bone of a right big toe discovered in 1961 in Olduvai Gorge in Tanzania—sets up a new signpost that not only clarifies the course of human evolution but also helps to guide those who speculate on the forces that converted predominantly quadrupedal animals into habitual bipeds.

Man's bipedal mode of walking seems potentially catastrophic because only the rhythmic forward movement of first one leg and then the other keeps him from falling flat on his face. Consider the sequence of events whenever a man sets out in pursuit of his center of gravity. A stride begins when the muscles of the calf relax and the walker's body sways forward (gravity supplying the energy needed to overcome the body's inertia). The sway places the center of body weight in front of the supporting pedestal normally formed by the two feet. As a result one or the other of the walker's legs must swing forward so that when his foot makes contact with the ground, the area of the supporting pedestal has been widened and the center of body weight once again rests safely within it. The pelvis plays an important role in this action: its degree of rotation determines the distance the swinging leg can move forward, and its muscles help to keep the body balanced while the leg is swinging.

At this point the "stance" leg—the leg still to the rear of the body's center of gravity—provides the propulsive force that drives the body forward. The walker applies this force by using muscular energy, pushing against the ground first with the ball of his foot and then with his big toe. The action constitutes the "push-off," which terminates the stance phase of the walking cycle. Once the stance foot leaves the ground, the walker's leg enters the starting, or "swing," phase of the cycle. As the leg swings forward it is able to clear the ground because it is bent at the hip, knee and ankle. This high-stepping action substantially reduces the leg's moment of inertia. Before making contact with the ground and ending the swing phase the leg straightens at the knee but remains bent at the ankle. As a result it is the

heel that strikes the ground first. The "heel strike" concludes the swing phase; as the body continues to move forward the leg once again enters the stance phase, during which the point of contact between foot and ground moves progressively nearer the toes. At the extreme end of the stance phase, as before, all the walker's propulsive thrust is delivered by the robust terminal bone of his big toe.

A complete walking cycle is considered to extend from the heel strike of one leg to the next heel strike of the same leg; it consists of the stance phase followed by the swing phase. The relative duration of the two phases depends on the cadence or speed of the walk. During normal walking the stance phase constitutes about 60 percent of the cycle and the swing phase 40 percent. Although

WALKING MAN, photographed by Eadweard Muybridge in 1884 during his studies of human and animal motion, exhibits the characteristic striding gait of the modern human.

the action of only one leg has been described in this account, the opposite leg obviously moves in a reciprocal fashion; when one leg is moving in the swing phase, the other leg is in its stance phase and keeps the body poised. Actually during normal walking the two phases overlap, so that both feet are on the ground at the same time for about 25 percent of the cycle. As walking speed increases, this period of double leg-support shortens.

Anyone who has watched other people walking and reflected a little on the process has noticed that the human stride demands both an up-and-down and a side-to-side displacement of the body. When two people walk side by side but out of step, the alternate bobbing of their heads makes it evident that the bodies undergo a vertical displacement with each stride. When two people walk in step but with opposite feet leading, they will sway first toward each other and then away in an equally graphic demonstration of the lateral displacement at each stride. When both displacements are plotted sequentially, a pair of low-amplitude sinusoidal curves appear, one in the vertical plane and the other in the horizontal [see illustrations on next page]. General observations of this kind were reduced to precise measurements during World War II when a group at the University of California at Berkeley led by H. D. Eberhart conducted a fundamental investigation of human walking in connection with requirements for the design of artificial legs. Eberhart and his colleagues found that a number of

functional determinants interacted to move the human body's center of gravity through space with a minimum expenditure of energy. In all they isolated six major elements related to hip, knee and foot movement that, working together, reduced both the amplitude of the two sine curves and the abruptness with which vertical and lateral changes in direction took place. If any one of these six elements was disturbed, an irregularity was injected into the normally smooth, undulating flow of walking, thereby producing a limp. What is more important, the irregularity brought about a measurable increase in the body's energy output during each step.

The Evidence of the Bones

What I have described in general and Eberhart's group studied in detail is the form of walking known as striding. It is characterized by the heel strike at the start of the stance phase and the push-off at its conclusion. Not all human walking is striding; when a man moves about slowly or walks on a slippery surface, he may take short steps in which both push-off and heel strike are absent. The foot is simply lifted from the ground at the end of the stance phase and set down flat at the end of the swing phase. The stride, however, is the essence of human bipedalism and the criterion by which the evolutionary status of a hominid walker must be judged. This being the case, it is illuminating to consider how the act of striding leaves its distinctive marks on the bones of the strider.

To take the pelvis first, there is a well-known clinical manifestation called Trendelenburg's sign that is regarded as evidence of hip disease in children. When a normal child stands on one leg, two muscles connecting that leg and the pelvis—the gluteus medius and the gluteus minimus—contract; this contraction, pulling on the pelvis, tilts it and holds it poised over the stance leg. When the hip is diseased, this mechanism fails to operate and the child shows a positive Trendelenburg's sign: the body tends to fall toward the unsupported side.

The same mechanism operates in walking, although not to the same degree. During the stance phase of the walking cycle, the same two gluteal muscles on the stance side brace the pelvis by cantilever action. Although actual tilting toward the stance side does not occur in normal walking, the action of the muscles in stabilizing the walker's hip is an essential component of the striding gait. Without this action the stride would become a slow, ungainly shuffle.

At the same time that the pelvis is stabilized in relation to the stance leg it also rotates to the unsupported side. This rotation, although small, has the effect of increasing the length of the stride. A familiar feature of the way women walk arises from this bit of anatomical mechanics. The difference in the proportions of the male and the female pelvis has the effect of slightly diminishing the range through which the female hip can move forward and back. Thus for a given length of stride women are obliged to rotate the pelvis through a greater

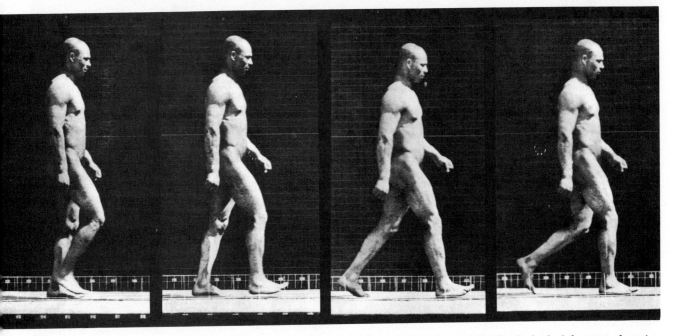

The free foot strikes the ground heel first and the body's weight is gradually transferred from heel to ball of foot as the opposite leg lifts and swings forward. Finally the heel of the stance foot rises and the leg's last contact with the ground is made with the big toe.

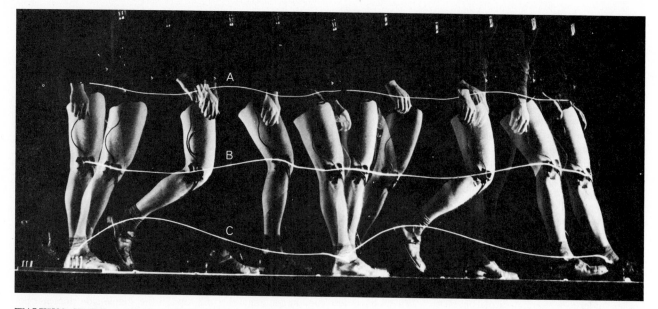

WALKING CYCLE extends from the heel strike of one leg to the next heel strike by the same leg. In the photograph, made by Gjon Mili in the course of a study aimed at improvement of artificial legs that he conducted for the U.S. Army, multiple exposures trace the progress of the right leg in the course of two strides. The ribbons of light allow analysis of the movement (see illustration below).

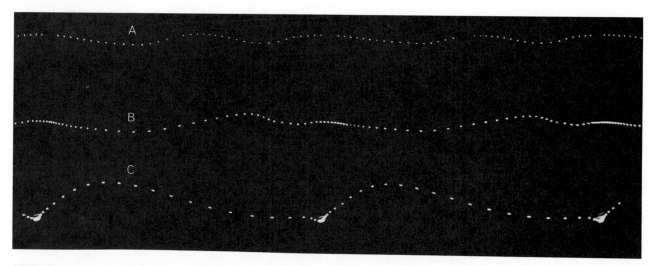

SINE CURVE described by the hip of a walking man was recorded on film by means of the experimental system illustrated above. An interrupter blade, passing in front of the camera lens at constant speed, broke the light from lamps attached to the walker into the three rows of dots. The speed of hip (a), knee (b) or ankle (c) during the stride is determined by measuring between the dots.

angle than men do. This secondary sexual characteristic has not lacked exploitation; at least in our culture female pelvic rotation has considerable erotogenic significance. What is more to the point in terms of human evolution is that both the rotation and the balancing of the pelvis leave unmistakable signs on the pelvic bone and on the femur: the leg bone that is joined to it. It is by a study of such signs that the walking capability of a fossil hominid can be judged.

Similar considerations apply to the foot. One way the role of the foot in walking can be studied is to record the vertical forces acting on each part of the foot while it is in contact with the ground

during the stance phase of the walking cycle. Many devices have been built for this purpose; one of them is the plastic pedograph. When the subject walks across the surface of the pedograph, a motion-picture camera simultaneously records the exact position of the foot in profile and the pattern of pressures on the surface. Pedograph analyses show that the initial contact between the striding leg and the ground is the heel strike. Because the foot is normally turned out slightly at the end of the swing phase of the walking cycle, the outer side of the back of the heel takes the brunt of the initial contact [see illustration on opposite page]. The outer side of the foot contin-

ues to support most of the pressure of the stance until a point about three-fifths of the way along the sole is reached. The weight of the body is then transferred to the ball of the foot and then to the big toe. In the penultimate stage of push-off the brunt of the pressure is under the toes, particularly the big toe. Finally, at the end of the stance phase, only the big toe is involved; it progressively loses contact with the ground and the final push-off is applied through its broad terminal bone.

The use of pedographs and similar apparatus provides precise evidence about the function of the foot in walking, but every physician knows that much the same information is recorded on the soles of everyone's shoes. Assuming that the shoes fit, their pattern of wear is a true record of the individual's habitual gait. The wear pattern will reveal a limp that one man is trying to hide, or unmask one that another man is trying to feign, perhaps to provide evidence for an insurance claim. In any case, just as the form of the pelvis and the femur can disclose the presence or absence of a striding gait, so can the form of the foot bones, particularly the form and proportions of the big-toe bones.

The Origins of Primate Bipedalism

Almost all primates can stand on their hind limbs, and many occasionally walk in this way. But our primate relatives are all, in a manner of speaking, amateurs; only man has taken up the business of bipedalism intensively. This raises two major questions. First, how did the basic postural adaptations that permit walking—occasional or habitual—arise among the primates? Second, what advantages did habitual bipedalism bestow on early man?

With regard to the first question, I have been concerned for some time with the anatomical proportions of all primates, not only man and the apes but also the monkeys and lower primate forms. Such consideration makes it possible to place the primates in natural groups according to their mode of locomotion. Not long ago I suggested a new group, and it is the only one that will concern us here. The group comprises primates with very long hind limbs and very short forelimbs. At about the same time my colleague Alan C. Walker, now at Makerere University College in Uganda, had begun a special study of the locomotion of living and fossil lemurs. Lemurs are among the most primitive offshoots of the basic primate stock. Early in Walker's studies he was struck by the frequency with which a posture best described as "vertical clinging" appeared in the day-to-day behavior of living lemurs. All the animals whose propensity for vertical clinging had been observed by Walker showed the same proportions—that is, long hind limbs and short forelimbs—I had proposed as forming a distinct locomotor group.

When Walker and I compared notes, we decided to define a hitherto unrecognized locomotor category among the primates that we named "vertical clinging and leaping," a term that includes both the animal's typical resting posture and the essential leaping component

in its locomotion. Since proposing this category a most interesting and important extension of the hypothesis has become apparent to us. Some of the earliest primate fossils known, preserved in sediments laid down during Eocene times and therefore as much as 50 million years old, are represented not only by skulls and jaws but also by a few limb bones. In their proportions and details most of these limb bones show the same characteristics that are displayed by the living members of our vertical-clinging-and-leaping group today. Not long ago Elwyn L. Simons of Yale University presented a reconstruction of the lemur-like North American Eocene primate *Smilodectes* walking along a tree branch in a quadrupedal position [see the article "The Early Relatives of Man," by Elwyn L. Simons, beginning on page 22]. Walker and I would prefer to see *Smilodectes* portrayed in the vertical clinging posture its anatomy unequivocally indicates. The fossil evidence, as far as it goes, suggests to us that vertical clinging and leaping was a major primate locomotor adaptation that took place some 50 million years ago. It may even have been the initial dynamic adaptation to tree life from which the subsequent locomotor patterns of all the living pri-

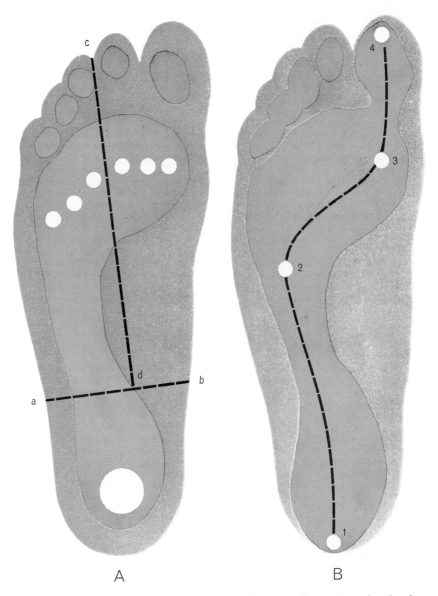

A   B

DISTRIBUTION OF WEIGHT in the human foot alters radically as action takes the place of rest. When motionless (*A*), the foot divides its static load (half of the body's total weight) between its heel and its ball along the axis *a–b*. The load on the ball of the foot is further divided equally on each side of the axis *c–d*. When striding (*B*), the load (all of the body's weight during part of each stride) is distributed dynamically from the first point of contact (1, *heel strike*) in a smooth flow via the first and fifth metatarsal bones (2, 3) that ends with a propulsive thrust (4, *push-off*) delivered by the terminal bone of the big toe.

mates, including man, have stemmed.

Walker and I are not alone in this view. In 1962 W. L. Straus, Jr., of Johns Hopkins University declared: "It can safely be assumed that primates early developed the mechanisms permitting maintenance of the trunk in the upright position.... Indeed, this tendency toward truncal erectness can be regarded as an essentially basic primate character." The central adaptations for erectness of the body, which have been retained in the majority of living primates, seem to have provided the necessary anatomical basis for the occasional bipedal behavior exhibited by today's monkeys and apes.

What we are concerned with here is the transition from a distant, hypothetical vertical-clinging ancestor to modern, bipedal man. The transition was almost

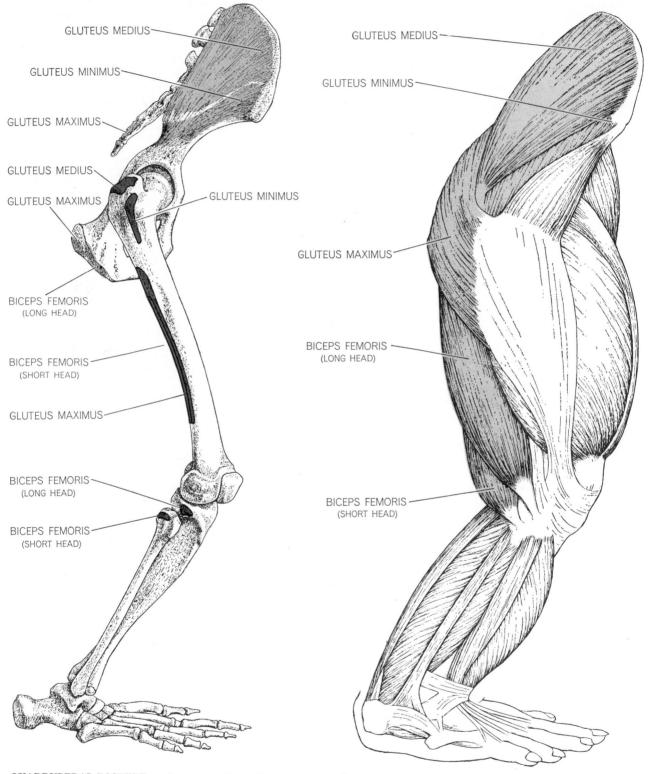

QUADRUPEDAL POSTURE needs two sets of muscles to act as the principal extensors of the hip. These are the gluteal group (the gluteus medius and minimus in particular), which connects the pelvis to the upper part of the femur, and the hamstring group. which connects the femur and the lower leg bones. Of these only the biceps femoris is shown in the gorilla musculature at right. The skeletal regions to which these muscles attach are shown in color at left. In most primates the gluteus maximus is quite small.

certainly marked by an intermediate quadrupedal stage. Possibly such Miocene fossil forms as *Proconsul*, a chimpanzee-like early primate from East Africa, represent such a stage. The structural adaptations necessary to convert a quadrupedal ape into a bipedal hominid are centered on the pelvis, the femur, the foot and the musculature associated with these bones. Among the nonhuman primates living today the pelvis and femur are adapted for four-footed walking; the functional relations between hipbones and thigh muscles are such that, when the animal attempts to assume a bipedal stance, the hip joint is subjected to a stress and the hip must be bent. To compensate for the resulting forward shift of the center of gravity, the knees must also be bent. In order to alter a bent-hip, bent-knee gait into

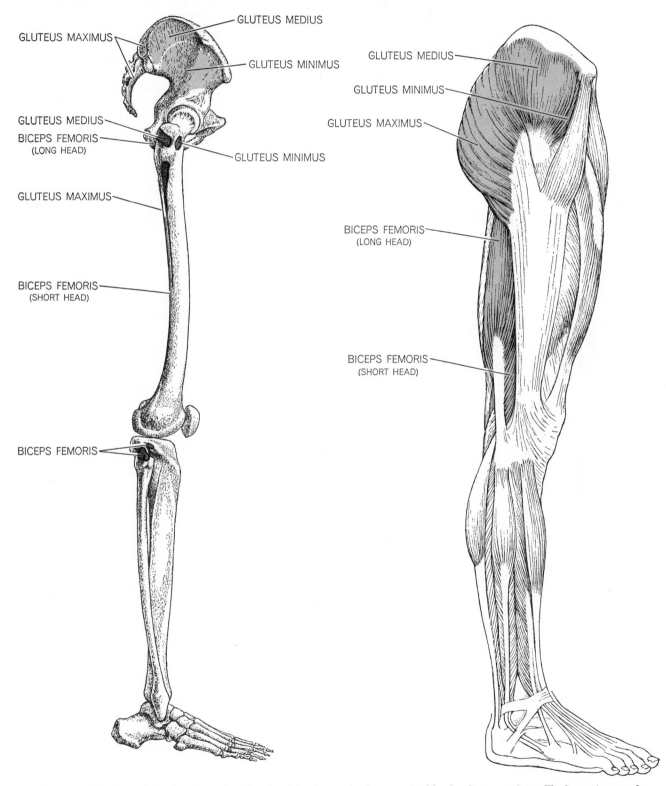

GLUTEUS MAXIMUS

GLUTEUS MEDIUS

GLUTEUS MINIMUS

GLUTEUS MEDIUS

GLUTEUS MINIMUS

GLUTEUS MEDIUS
BICEPS FEMORIS
(LONG HEAD)

GLUTEUS MINIMUS

GLUTEUS MAXIMUS

GLUTEUS MEDIUS

GLUTEUS MINIMUS

GLUTEUS MAXIMUS

BICEPS FEMORIS
(SHORT HEAD)

BICEPS FEMORIS
(LONG HEAD)

BICEPS FEMORIS

BICEPS FEMORIS
(SHORT HEAD)

BIPEDAL POSTURE brings a reversal in the roles played by the same pelvic and femoral muscles. Gluteus medius and gluteus minimus have changed from extensors to abductors and the function of extending the trunk, required when a biped runs or climbs, has been assumed by the gluteus maximus. The hamstring muscles, in turn, now act mainly as stabilizers and extensors of the hip. At right are the muscles as they appear in man; the skeletal regions to which their upper and lower ends attach are shown in color at left.

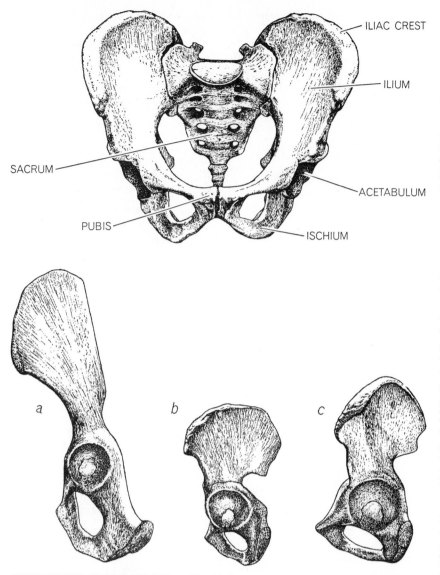

ILIAC CREST

ILIUM

SACRUM

ACETABULUM

PUBIS

ISCHIUM

a   b   c

COMPONENTS OF THE PELVIS are identified at top; the bones are those of the human pelvis. Below, ilium and ischium of a gorilla (*a*), of *Australopithecus* (*b*) and of modern man (*c*) are seen from the side (the front is to the left in each instance). The ischium of *Australopithecus* is longer than man's; this almost certainly kept the early hominid from striding in the manner of *Homo sapiens*. Instead the gait was probably a kind of jog trot.

prisingly unimportant role in man's ability to stand, or even to walk on a level surface. In standing, for example, the principal stabilizing and extending agents are the muscles of the hamstring group. In walking on the level the gluteus maximus is so little involved that even when it is paralyzed a man's stride is virtually unimpaired. The gluteus maximus comes into its own in man when power is needed to give the hip joint more play for such activities as running, walking up a steep slope or climbing stairs [see illustration on page 58]. Its chief function in these circumstances is to correct any tendency for the human trunk to jackknife on the legs.

Because the gluteus maximus has such a specialized role I believe, in contrast to Washburn's view, that it did not assume its present form until late in the evolution of the striding gait. Rather than being the initial adaptation, this muscle's enlargement and present function appear to me far more likely to have been one of the ultimate refinements of human walking. I am in agreement with Washburn, however, when he states that changes in the ilium, or upper pelvis, would have preceded changes in the ischium, or lower pelvis [see the article "Tools and Human Evolution," by Sherwood L. Washburn, beginning on page 9]. The primary adaptation would probably have involved a forward curvature of the vertebral column in the lumbar region. Accompanying this change would have been a broadening and a forward rotation of the iliac portions of the pelvis. Together these early adaptations provide the structural basis for improving the posture of the trunk.

Assuming that we have now given at least a tentative answer to the question of how man's bipedal posture evolved, there remains to be answered the question of why. What were the advantages of habitual bipedalism? Noting the comparative energy demands of various gaits, Washburn points out that human walking is primarily an adaptation for covering long distances economically. To go a long way with a minimum of effort is an asset to a hunter; it seems plausible that evolutionary selection for hunting behavior in man was responsible for the rapid development of striding anatomy. Gordon W. Hewes of the University of Colorado suggests a possible incentive that, acting as an agent of natural selection, could have prompted the quadrupedal ancestors of man to adopt a two-footed gait. In Hewes's view the principal advantage of bipedalism over quadrupedalism would be the free-

man's erect, striding walk, a number of anatomical changes must occur. These include an elongation of the hind limbs with respect to the forelimbs, a shortening and broadening of the pelvis, adjustments of the musculature of the hip (in order to stabilize the trunk during the act of walking upright), a straightening of both hip and knee and considerable reshaping of the foot.

Which of these changes can be considered to be primary and which secondary is still a matter that needs elucidation. Sherwood L. Washburn of the University of California at Berkeley has expressed the view that the change from four-footed to two-footed posture was initiated by a modification in the form and function of the gluteus maximus, a thigh muscle that is powerfully

developed in man but weakly developed in monkeys and apes [see illustrations on preceding two pages]. In a quadrupedal primate the principal extensors of the trunk are the "hamstring" muscles and the two upper-leg muscles I have already mentioned: the gluteus medius and gluteus minimus. In man these two muscles bear a different relation to the pelvis, in terms of both position and function. In technical terms they have become abductor muscles of the trunk rather than extensor muscles of the leg. It is this that enables them to play a critical part in stabilizing the pelvis in the course of striding. In man the extensor function of these two gluteal muscles has been taken over by a third, the gluteus maximus. This muscle, insignificant in other primates, plays a sur-

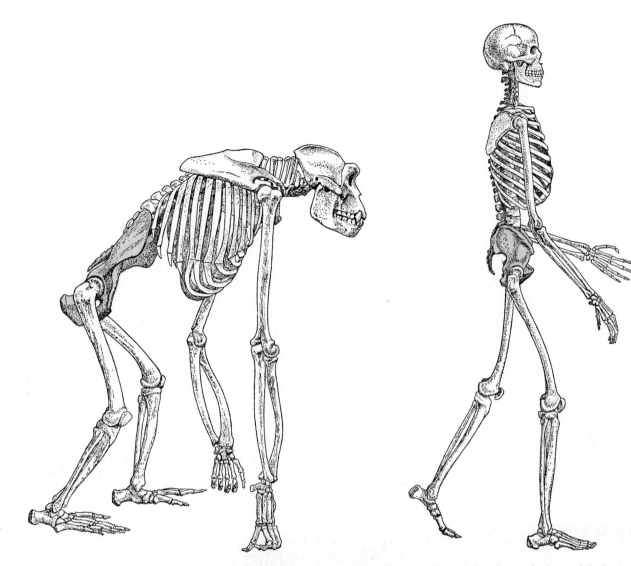

SHAPE AND ORIENTATION of the pelvis in the gorilla and in man reflect the postural differences between quadrupedal and bipedal locomotion. The ischium in the gorilla is long, the ilium extends to the side and the whole pelvis is tilted toward the horizontal (*see illustration on opposite page*). In man the ischium is much shorter, the broad ilium extends forward and the pelvis is vertical.

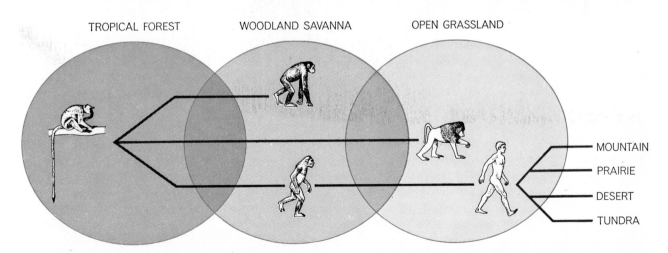

TROPICAL FOREST   WOODLAND SAVANNA   OPEN GRASSLAND

MOUNTAIN

PRAIRIE

DESERT

TUNDRA

ECOLOGICAL PATHWAY to man's eventual mastery of all environments begins (*left*) with a quadrupedal primate ancestor living in tropical forest more than 20 million years ago. During Miocene times mountain-building produced new environments. One, a transition zone between forest and grassland, has been exploited by three groups of primates. Some, for example the chimpanzees, have only recently entered this woodland savanna. Both the newly bipedal hominids and some ground-living quadrupedal monkeys, however, moved beyond the transition zone into open grassland. The quadrupeds, for example the baboons, remained there. On the other hand, the forces of natural selection in the new setting favored the bipedal hominid hunters' adaptation of the striding gait typical of man. Once this adaptation developed, man went on to conquer most of the earth's environments.

ing of the hands, so that food could be carried readily from one place to another for later consumption. To assess the significance of such factors as survival mechanisms it behooves us to review briefly the ecological situation in which our prehuman ancestors found themselves in Miocene times, between 15 and 25 million years ago.

The Miocene Environment

During the Miocene epoch the world-wide mountain-building activity of middle Tertiary times was in full swing. Many parts of the earth, including the region of East Africa where primates of the genus *Proconsul* were living, were being faulted and uplifted to form such mountain zones as the Alps, the Himalayas, the Andes and the Rockies. Massive faulting in Africa gave rise to one of the earth's major geological features: the Rift Valley, which extends 5,000 miles from Tanzania across East Africa to Israel and the Dead Sea. A string of lakes lies along the floor of the Rift Valley like giant stepping-stones. On their shores in Miocene times lived a fantastically rich fauna, inhabitants of the forest and of a new ecological niche—the grassy savanna.

These grasslands of the Miocene were the domain of new forms of vegetation that in many parts of the world had taken the place of rain forest, the dominant form of vegetation in the Eocene and the Oligocene. The savanna offered new evolutionary opportunities to a variety of mammals, including the expanding population of primates in the rapidly shrinking forest. A few primates—the ancestors of man and probably also the ancestors of the living baboons—evidently reacted to the challenge of the new environment.

The savanna, however, was no Eldorado. The problems facing the early hominids in the open grassland were immense. The forest foods to which they were accustomed were hard to come by; the danger of attack by predators was immeasurably increased. If, on top of everything else, the ancestral hominids of Miocene times were in the process of converting from quadrupedalism to bipedalism, it is difficult to conceive of any advantage in bipedalism that could have compensated for the added hazards of life in the open grassland. Consideration of the drawbacks of savanna living has led me to a conclusion contrary to the one generally accepted: I doubt that the advent of bipedalism took place in this environment. An environment neglected by scholars but one far better

suited for the origin of man is the woodland-savanna, which is neither high forest nor open grassland. Today this halfway-house niche is occupied by many primates, for example the vervet monkey and some chimpanzees. It has enough trees to provide forest foods and ready escape from predators. At the same time its open grassy spaces are arenas in which new locomotor adaptations can be practiced and new foods can be sampled. In short, the woodland-savanna provides an ideal nursery for evolving hominids, combining the challenge and incentive of the open grassland with much of the security of the forest. It was probably in this transitional environment that man's ancestors learned to walk on two legs. In all likelihood, however, they only learned to stride when they later moved into the open savanna.

Moving forward many millions of years from Miocene to Pleistocene times, we come to man's most immediate hominid precursor: *Australopithecus*. A large consortium of authorities agrees that the shape of the pelvis in *Australopithecus* fossils indicates that these hominids were habitually bipedal, although not to the degree of perfection exhibited by modern man. A few anatomists, fighting a rearguard action, contend that on the contrary the pelvis of *Australopithecus*

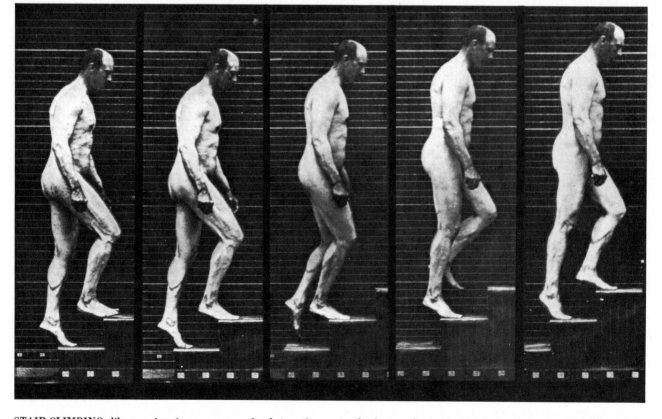

STAIR-CLIMBING, like running, is a movement that brings the human gluteus maximus into play. Acting as an extensor of the trunk, the muscle counteracts any tendency for the body to jackknife over the legs. Photographs are from Muybridge's collection.

shows that these hominids were predominantly quadrupedal. I belong to the first school but, as I have been at some pains to emphasize in the past, the kind of upright walking practiced by *Australopithecus* should not be equated with man's heel-and-toe, striding gait.

From Bipedalist to Strider

The stride, although it was not necessarily habitual among the earliest true men, is nevertheless the quintessence of the human locomotor achievement. Among other things, striding involves extension of the leg to a position behind the vertical axis of the spinal column. The degree of extension needed can only be achieved if the ischium of the pelvis is short. But the ischium of *Australopithecus* is long, almost as long as the ischium of an ape [*see illustration on page 42*]. Moreover, it has been shown that in man the gluteus medius and the gluteus minimus are prime movers in stabilizing the pelvis during each stride; in *Australopithecus* this stabilizing mechanism is imperfectly evolved. The combination of both deficiencies almost entirely precludes the possibility that these hominids possessed a striding gait. For *Australopithecus* walking was something of a jog trot. These hominids must have covered the ground with quick, rather short steps, with their knees and hips slightly bent; the prolonged stance phase of the fully human gait must surely have been absent.

Compared with man's stride, therefore, the gait of *Australopithecus* is physiologically inefficient. It calls for a disproportionately high output of energy; indeed, *Australopithecus* probably found long-distance bipedal travel impossible. A natural question arises in this connection. Could the greater energy requirement have led these early representatives of the human family to alter their diet in the direction of an increased reliance on high-energy foodstuffs, such as the flesh of other animals?

The pelvis of *Australopithecus* bears evidence that this hominid walker could scarcely have been a strider. Let us now turn to the foot of what many of us believe is a more advanced hominid. In 1960 L. S. B. Leakey and his wife Mary unearthed most of the bones of this foot in the lower strata at Olduvai Gorge known collectively as Bed I, which are about 1.75 million years old. The bones formed part of a fossil assemblage that has been designated by the Leakeys, by Philip Tobias of the University of the Witwatersrand and by me as possibly the earliest-known species of man: *Homo*

habilis. The foot was complete except for the back of the heel and the terminal bones of the toes; its surviving components were assembled and studied by me and Michael Day, one of my colleagues at the Unit of Primatology and Human Evolution of the Royal Free Hospital School of Medicine in London. On the basis of functional analysis the resem-

blance to the foot of modern man is close, although differing in a few minor particulars. Perhaps the most significant point of resemblance is that the stout basal bone of the big toe lies alongside the other toes [*see upper illustration on next page*]. This is an essentially human characteristic; in apes and monkeys the big toe is not exceptionally robust and

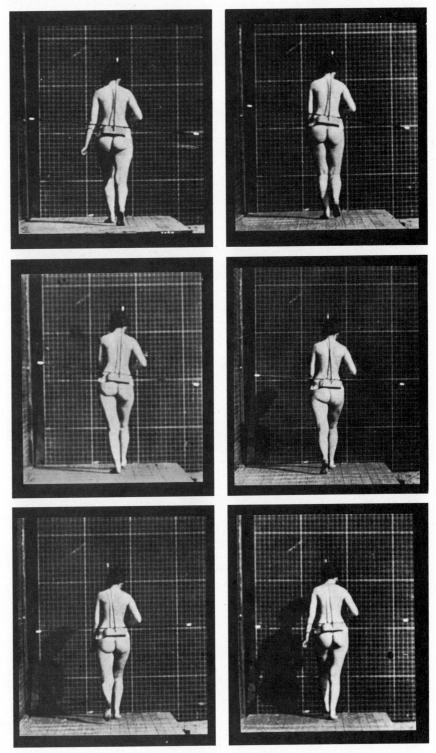

PELVIC ROTATION of the human female is exaggerated compared with that of a male taking a stride of equal length because the two sexes differ in pelvic anatomy. Muybridge noted the phenomenon, using a pole with whitened ends to record the pelvic oscillations.

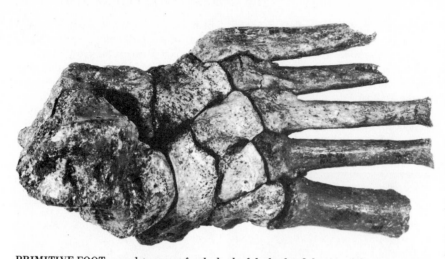

PRIMITIVE FOOT, complete except for the back of the heel and the tips of the toes, was un-earthed from the lower level at Olduvai Gorge in Tanzania. Attributed to a very early homi-nid, *Homo habilis*, by its discoverer, L. S. B. Leakey, it is about 1.75 million years old. Its appearance suggests that the possessor was a habitual biped. Absence of the terminal bones of the toes, however, leaves open the question of whether the possessor walked with a stride.

BIG-TOE BONE, also discovered at Olduvai Gorge, is considerably younger than the foot bones in the top illustration but still probably more than a million years old. It is the toe's terminal bone (*bottom view at left, top view at right*) and bore the thrust of its possessor's push-off with each swing of the right leg. The tilting and twisting of the head of the bone in relation to the shaft is unequivocal evidence that its possessor walked with a modern stride.

diverges widely from the other toes. The foot bones, therefore, give evidence that this early hominid species was habitually bipedal. In the absence of the terminal bones of the toes, however, there was no certainty that *Homo habilis* walked with a striding gait.

Then in 1961, in a somewhat higher stratum at Olduvai Gorge (and thus in a slightly younger geological formation), a single bone came to light in an area oth-erwise barren of human bones. This fos-sil is the big-toe bone I mentioned at the beginning of this article [*see lower illus-tration at left*]. Its head is both tilted and twisted with respect to its shaft, charac-teristics that are found only in modern man and that can with assurance be cor-related with a striding gait. Day has re-cently completed a dimensional analysis of the bone, using a multivariate statisti-cal technique. He is able to show that the fossil is unquestionably human in form.

There is no evidence to link the big-toe bone specifically to either of the two recognized hominids whose fossil re-mains have been recovered from Bed I at Olduvai: *Homo habilis* and *Zinjan-thropus boisei*. Thus the owner of the toe remains unknown, at least for the pres-ent. Nonetheless, one thing is made cer-tain by the discovery. We now know that in East Africa more than a million years ago there existed a creature whose mode of locomotion was essentially hu-man.

The Hominids
of East Turkana

by Alan Walker and Richard E. F. Leakey
August 1978

*This region on the shore of Lake Turkana in
northeastern Kenya is a treasure trove of fossils of
early members of the genus Homo and their close
relatives dating back 1.5 million years and more*

The continent of Africa is rich in the fossilized remains of extinct mammals, and one of the richest repositories of such remains is located in Kenya, near the border with Ethiopia. The first European to explore the area was a 19th-century geographer, Count Samuel Teleki, who reached the forbidding eastern shore of an unmapped brackish lake there early in 1888. Exercising the explorer's prerogative, he named the 2,500-square-mile body of water after the Austro-Hungarian emperor Franz Josef's son and heir, the archduke Rudolf (who within a year had committed suicide in the notorious Mayerling episode). Teleki's party evidently passed a major landmark on the shore of the lake, the Koobi Fora promontory, during the last week in March. They took notes on the local geology and even collected a few fossil shells, but they missed the mammal remains.

Teleki's name has faded into history, and even the name of the lake has now been changed by the government of Kenya from Lake Rudolf to Lake Turkana. Yet today the Koobi Fora area is world-famous among students of early man. Its mammalian fossils include the partial remains of some 150 individual hominids, early relatives of modern man. They represent the most abundant and varied assemblage of early hominid fossils found so far anywhere in the world.

The fossil beds of East Turkana (formerly East Rudolf) might have been found at any time after Teleki's reconnaissance. It was not until 1967, however, that the deposits came to notice. At that time an international group was authorized by the government of Ethiopia to study the geology of a remote southern corner of the country: the valley of the Omo River, a tributary of Lake Turkana. Erosion in the area has exposed sedimentary strata extending backward in time from the Pleistocene to the Pliocene, that is, from about one million to about four million years ago.

Supplies going to the Omo camps were flown over the East Turkana area; on one such trip one of us (Leakey) noticed that part of the terrain consisted of sedimentary beds that had been dissected by streams and that appeared to be potentially fossil-bearing. A brief survey afterward by helicopter revealed that the exposed sediments contained not only mammalian fossils but also stone tools. This reconnaissance was followed up in 1968 by an expedition to the vast, hot and inhospitable area. Out of a total area of several thousand square kilometers the expedition located some 800 square kilometers of fossil-bearing sediments, mainly in the vicinity of Koobi Fora, Ileret and to the south at Allia Bay. The expedition also found the fossil remains of many kinds of mammals, most of them beautifully preserved. Only in the category of hominids were the finds disappointing: the total was only three jaws, all of them badly weathered. Nevertheless, the overall richness of the fossil deposits made it clear that further prospecting would be worthwhile.

The large task of establishing the geological context of both the fossils and the stone tools discovered in East Turkana began in 1969. The work that season was highlighted by the excavation of the first stone tools to be found in stratified sequences there and by the discovery of two skulls of early hominids. It was with these discoveries that the enormous importance of the area for the study of human evolution began to be recognized.

A formal organization was established: the Koobi Fora Research Project. The project operates under the joint leadership of one of us (Leakey) and Glynn Isaac of the University of California at Berkeley. In the years since it was founded the project has brought together workers from many countries who represent many different disciplines: geology, geophysics, paleontology, anatomy, archaeology, ecology and taphonomy. (Taphonomy is a new discipline concerned with the study of the processes that convert living plant and animal communities into collections of fossils.) The interaction of specialists has become a particular strength of the project as the workers have become increasingly aware of the particular outlook (and also the limitations) of fields other than their own.

The project area extends from the Kenya-Ethiopia border on the north to a point south of Allia Bay where the land surface is of volcanic origin. The western boundary is the lakeshore and the eastern is marked by another volcanic outcropping. The promontory of Koobi Fora itself, a spit of land that extends a few hundred meters into the lake, is the site of our base camp. Each of the three principal areas of fossil-bearing sediments has its own natural boundaries; when these areas are seen from the air, they show up as pale patches among the darker volcanic terrain. For reference purposes they have been divided into smaller units that are identified by number on the project maps and that are readily distinguished in the field by their vegetation, dry rivers and the like.

In studies such as these it is of paramount importance to develop a chronological framework that allows the fossil finds to be placed in their correct relative positions. The construction of such a framework is the responsibility of the project geologists and geophysicists. The geology of East Turkana is straightforward in its broad outlines but extremely complex in many of its local details. Among the factors responsible for its complexity are abrupt lateral shifts in the composition of the exposed sedimentary strata, discontinuities, faulting that involves rather small displacements and above all the absence from many of the sediments of volcanic tuff: layers of ash that play a key role in correlating the strata.

The difficult work of geological mapping and stratigraphic correlation has been carried out by many of our colleagues but mainly by Bruce Bowen of Iowa State University, working in collaboration with Ian Findlater of the International Louis Leakey Memorial Institute for African Prehistory in Nairobi, Kay Behrensmeyer of Yale University and Carl F. Vondra of Iowa State.

FOSSIL-RICH AREAS in East Turkana appear as pale patches in a satellite image of the region. The deltas of the Omo River, which rises in Ethiopia, appear at the northern end of Lake Turkana (formerly Lake Rudolf). The short, narrow promontory that juts out from the eastern shore of the lake is Koobi Fora, where the base camp for project fieldworkers is situated. Imagery is from the satellite *Landsat 2*.

The complexities have nonetheless prevented the accurate placement of some of the most important early hominid fossils in the stratigraphic framework the geologists established. In such cases we have provisionally assigned the specimens to temporal positions on the basis of criteria other than stratigraphic ones.

The fossil-rich sediments are underlain by older rocks of volcanic origin. The sediments themselves are of various kinds, laid down in such different ancient environments as stream channels and their associated floodplains, lake bottoms, stream deltas and former lakeshores. For the most part the strata dip gently toward the present Lake Turkana. In the past, extensions of various deltas and coastal plains frequently built out westward into the former lake basin; these intrusions alternated with periods when the lake waters intruded eastward. The result is a complex interdigitation of lake sediments and stream sediments.

The major basis for correlating the various strata is the presence of distinctive strata consisting of tuffs; the volcanic material has periodically washed into the lakeshore basin from the terrain to the north and to the east. Some of the tuff beds are widespread and some are not. The uncertainties of correlation between tuff layers in different locations are greatest in the Koobi Fora and Allia Bay areas; these are the areas farthest from the erosional sources of the volcanic ash. At the same time volcanic rocks can be dated by means of isotope measurements, which makes the ash strata particularly important.

Jack Miller of the University of Cambridge and Frank J. Fitch of Birkbeck College in London have conducted most of the dating studies of the tuff (based on the decay of radioactive potassium into argon). Independent chronological data are also available from studies of the magnetic orientation of some particles in the volcanic ash and studies of fission tracks in bits of zircon in the ash. The various measurements are by no means unequivocal, but it can be stated as a generality that the location of a fossil find below one such layer of tuff and above another at least establishes a relative chronological position for the fossil even if its precise age remains in doubt.

There are five principal tuff marker layers. The earliest, which provides a boundary between the Kubi Algi sediments below it and the Koobi Fora sediments above it, is the Surgaei tuff. The next layer of tuff divides the lower member of the Koobi Fora sedimentary formation approximately in half; this is the Tulu Bor tuff, 3.2 million years old. The next layer, the KBS tuff, is named for the exposure where it was first recognized: the Kay Behrensmeyer site. It marks the boundary between the lower and upper members of the Koobi Fora Formation, and its exact age is debated.

KOOBI FORA REGION of East Turkana lies near the border between Kenya and Ethiopia. The fossil material collected there is brought back to Nairobi for preparation and analysis.

In 1970 Miller and Fitch ran a potassium/argon analysis of the KBS tuff that showed it to be 2.61 (±.26) million years old. They have recently made new calculations indicating that the tuff is 2.42 (±.01) million years old. At the same time an age determination based on other KBS samples, done by Garniss H. Curtis of the University of California at Berkeley, yields two much younger readings: 1.82 (±.04) million and 1.60 (±.05) million years.

Above the anomalous KBS tuff the next marker layer is the Okote tuff, which divides the upper member of the Koobi Fora sediments approximately in half. The Okote tuff is between 1.6 and 1.5 million years old. The fifth and uppermost marker layer, roughly indicating the boundary between the Koobi Fora sediments and the overlying Guomde sediments, is the Chari-Karari tuff; it is between 1.3 and 1.2 million years old.

We present the stratigraphy in such detail because most of the early hominid fossils discovered thus far in East Turkana are sandwiched between the Tulu Bor tuff at the most ancient end of the geological column and the Chari-Karari tuff at the most recent end. Twenty-six specimens, including a remarkable skull unearthed in 1972, designated KNM-ER 1470, come from sediments that lie below the KBS tuff. (The designation is an abbreviation of the formal accession number: Kenya National Museum–East Rudolf No. 1470.) Another 34 specimens, including several skulls, come

from sediments that lie above the KBS tuff but below the Okote tuff. Unfortunately for those interested in measuring the rates of evolutionary change in hominid lineages, the difference between the oldest and the youngest proposed KBS-tuff dates is some 1.3 million years.

That span of time far exceeds the one allotted to the whole of human evolution not many years ago. Even by today's standards the KBS-tuff discrepancy is enough to allow uncomfortably different evolutionary rates for various hypothetical hominid lineages. So far evidence of other kinds has not resolved the issue. For example, our colleagues John Harris of the Louis Leakey Memorial Institute and Timothy White of the University of California at Berkeley, who have conducted a detailed study of the evolution of pigs throughout Africa, suggest that the more recent date for the KBS tuff would best suit their fossil data. At the same time the fission-track studies of zircons from the KBS tuff indicate that the older dates are correct. For the time being we must accept the fact that the KBS tuff is either about 2.5 million years old or somewhere between 1.8 and 1.6 million years old.

Relative and absolute chronology apart, other kinds of investigation are increasing our knowledge of the different environments that were inhabited by the East Turkana hominids. For example, most of the hominid specimens can in general be placed either in the genus *Australopithecus* or in the genus *Homo*. Behrensmeyer, Findlater and Bowen

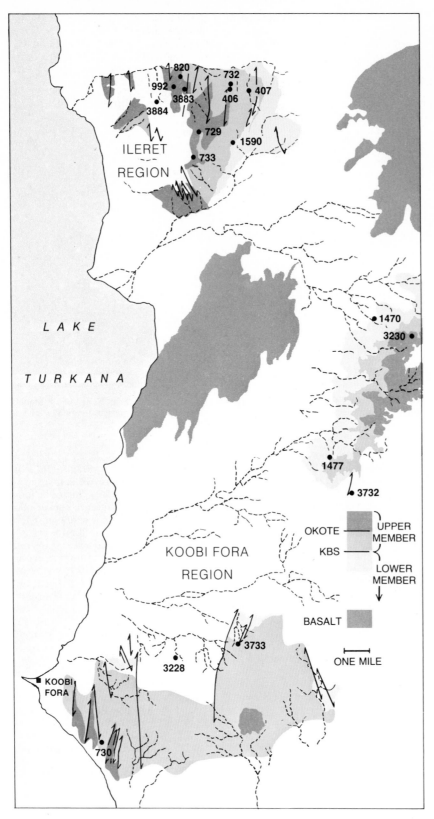

EAST TURKANA FOSSILS are found in a complex interdigitation of sedimentary rocks, some of lakeshore and delta origin and others of streambed origin, that were laid down during alternating periods of lake transgression and land buildup. The major geological marker layers are beds of volcanic tuffs that have been washed from the east and north across parts of the region. Two of the three main fossil-bearing areas are the Ileret region (*top*) and the Koobi Fora region (*bottom*). Allia Bay is not shown. The map is based on the work of Ian Findlater; numbers identify some of the hominid-fossil finds in both regions. Geological faults throughout the region are identified by conventional symbols; also identified (*color*) are sedimentary strata of the upper member and part of the lower member of the Koobi Fora Formation. These strata are separated by two volcanic-tuff marker beds: the Okote complex and the KBS complex.

are engaged in microstratigraphic studies that have enabled Behrensmeyer to associate many of the specimens with a specific environment of sedimentary burial. Preliminary analyses indicate that the specimens identified as *Homo* were fossilized more commonly in lake-margin sediments than in stream sediments whereas the specimens identified as *Australopithecus* are equally common in both sedimentary environments. Facts such as these promise to be of great help in reconstructing the lives of early hominids. In this instance the chance that an organism will be buried near where it spends most of its time is greater than the chance that it will be buried farther away. Thus Behrensmeyer has hypothesized that in this region of Africa early *Homo* exhibited a preference for living on the lakeshore.

Because of the unusual circumstances in this badlands region it will be useful to describe how the hominid fossils have been collected. The initial process is one of surface prospecting. The Kenyan prospecting team is led by Bwana Kimeu Kimeu; its job is to locate areas where natural erosion has left scatterings of mammalian bones and teeth exposed on the arid surface of the sedimentary beds. Kimeu is highly skilled at recognizing even fragmentary bits of hominid bone in the general bone litter present in such exposures.

Once the presence of a hominid fossil is established by the prospectors one of the project geologists determines its position with respect to the local stratigraphic section and records the location. Thereafter one of two procedures is generally followed. If the bone fragment has been washed completely free from the sedimentary matrix that held it, the practice is to scrape down and sieve the entire surrounding surface area in the hope of recovering additional fragments. As the scraping is done a watchful eye is kept for fragments that might still be in situ, that is, partly or entirely embedded in the rock.

If the initial discovery is a fossil fragment in situ, the procedure is different. Excavation is begun on a near-microscopic scale, the tools being dental picks and brushes. The Turkana hominid fossils are often so little mineralized that a preservative must be applied to the bone as excavation progresses in order to keep it from fragmenting further. Indeed, sometimes the preservative fluid must be applied with painstaking care because the impact of a falling drop can cause breakage.

After excavation each site is marked by a concrete post inscribed with an accession number provided by the Kenya National Museum. The next task, usually undertaken in the project laboratory in Nairobi, is piecing together the specimens. This is rather like doing a three-dimensional jigsaw puzzle with many of the pieces missing and no picture on the box. Any adhering matrix is now removed under the microscope, most often with an air-powered miniature jackhammer. (Cleaning with acid, a common laboratory method, is out of the question because the fossil bone is less resistant to the acid than the matrix.)

Finally, the hardened pieces of bone are reconstructed to the extent possible by gluing adjacent fragments together.

Can the East Turkana collection be considered representative of the hominid populations that occupied the area more than a million years ago? Taphonomy, the relatively new discipline that attempts to define the processes whereby communities of plants and animals do or do not become preserved as fossils, is beginning to provide some helpful answers to the question. The biases that affect fossil samples are many. For example, circumstances may result in the preservation of only some parts of

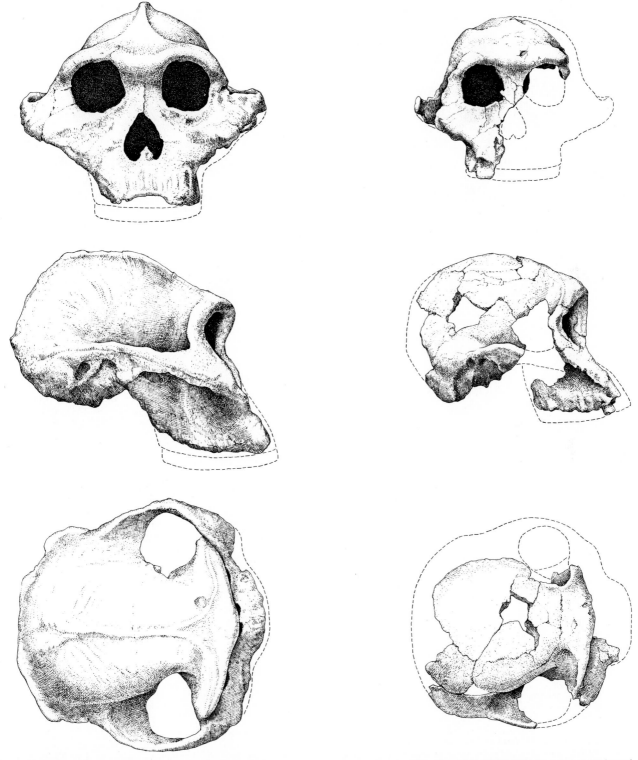

HOMINID FOSSILS found early in the process of collecting in East Turkana are illustrated. At the left is Kenya National Museum–East Rudolf (KNM-ER) accession No. 406, a robust cranium with well-preserved facial bones. At right is KNM-ER 732, a fragmented cranium that has little of the face preserved and is less robust than KNM-ER 406. Both specimens are placed in the genus *Australopithecus*.

certain individuals. Or a particular specimen may be severely deformed by pressure during its long burial. One bias that is easy to recognize in the East Turkana collection is a disproportionate number of lower jaws of the early hominid *Australopithecus robustus*. This hominid had powerful jaws and unusually large teeth; its lower jawbone is particularly massive. The relative abundance of these jaws and teeth in the East Turkana sediments probably results more from their mechanical strength, and thus their enhanced ability to survive fossilization, than from any preponderance of *A. robustus* individuals in the population.

Another example of bias in the hominid-fossil collection is the disproportionate representation of different parts of the skeleton. Teeth are by far the hardest parts, and so it is not surprising to find that teeth account for the largest fraction of the East Turkana sample. In contrast, vertebrae and hand and foot

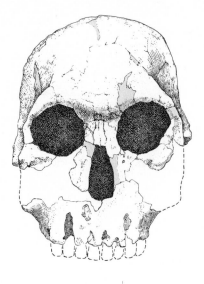

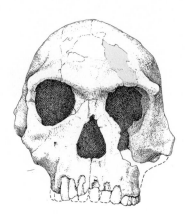

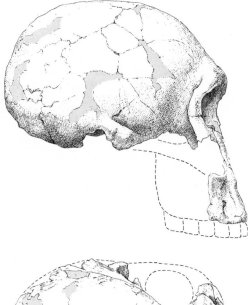

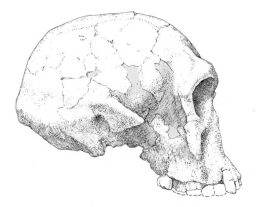

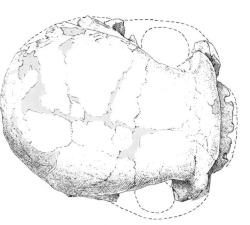

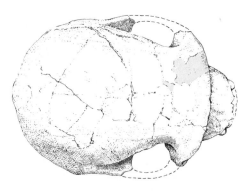

HOMINID FOSSILS OF DIFFERENT AGE are KNM-ER 1470, at left, and KNM-ER 1813, at right. The first cranium comes from the lower member of the Koobi Fora Formation; it cannot be less than 1.6 million years old and may be more than 2.5 million years old. It has a cranial capacity of about 775 cubic centimeters, compared with the *Australopithecus* average of about 500 c.c. The second cranium is provisionally assigned to the upper member of the Koobi Fora Formation, suggesting that it is no less than 1.2 million and may be more than 1.6 million years old. Its cranial capacity is about 500 c.c. It resembles other African hominid fossils 1.5 to two million years old.

bones are rarely found. Can this bias be attributed to the destructive processes associated with burial and exposure alone? It seems only logical to take into account a third process: carnivore and scavenger feeding on the hominid bodies before the sediments covered them.

What fraction of the hominid population in East Turkana more than a million years ago might our fossil collection represent? The answer, based on modern population studies of wild dogs and baboons, is that the fraction is extremely small. Assuming an appropriate interval between generations, if the hominid population density was low, as it is among living wild dogs, the collection represents two ten-thousandths, or .02 percent, of the original population. If the hominid population density was high, as it is among baboons, the fraction is very much smaller: two ten-millionths, or .00002 percent. If we ask further what fraction of the ancient popu

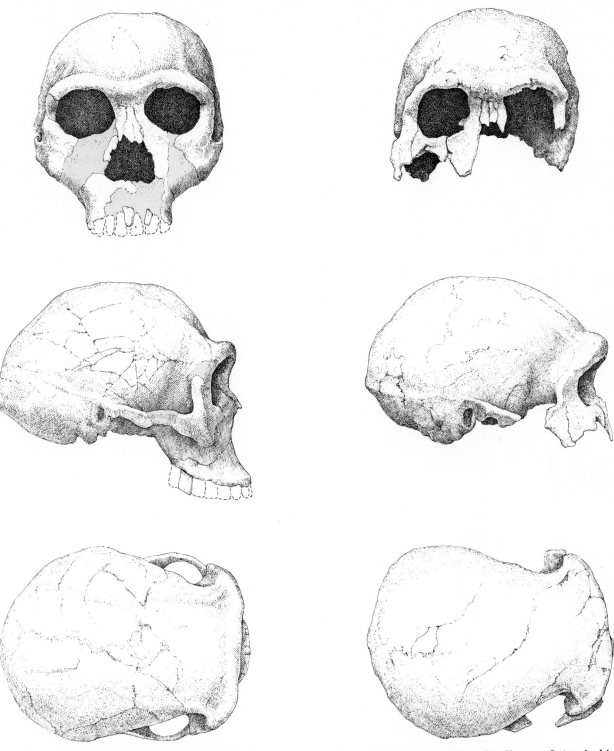

TWO SKULLS OF THE GENUS HOMO from the East Turkana fossil beds represent the early human species *Homo erectus*, a fossil hominid first discovered in Java and China. These are KNM-ER 3733 and KNM-ER 3883. Both are more than 1.5 million years old, which makes them a million years older than the specimens from China. Their great age strongly suggests that *H. erectus* first evolved in Africa. KNM-ER 3733 has a cranial capacity of about 850 c.c.; the cranial capacity of KNM-ER 3883 has not yet been measured but is probably about the same. Some specimens of *H. erectus* from Africa and Java have cranial capacities that are greater than 1,000 c.c.

lation is represented by the relatively complete skulls in the collection, it may be smaller still: it is between a hundred-thousandth and a hundred-millionth of the total. The second figure is the equivalent of someone's selecting two individuals at random to represent the entire population of the U.S. today. It is on this small sample that our hypotheses concerning hominid evolution must be based.

In developing such hypotheses we must keep in mind a number of fundamental questions. One question is: How many different species of early hominids were there in East Turkana? Another question is whether those species, whatever their number, existed over long periods of time or were replaced by other species. Again, do any or all of the species show signs of evolution during this interval of perhaps 1.5 million years or perhaps only 700,000 years? If there was any evolutionary change, what was its nature? How did the early hominids feed themselves? Were they relatively low-energy herbivores or relatively high-energy omnivores? If indeed several species were present at the same time, did each occupy a distinct ecological niche? What kept the niches separate?

The questions do not end here. Other questions, more specifically anatomical, also call for answers. Do the hominid fossils possess any morphological attributes that might be correlated with the archaeological record of tool use and scavenger-hunter behavior in the area? For example, can we detect any evidence of significant brain evolution during the period? Are there any morphological changes suggestive of altered patterns of locomotion or hand use that might shed light on the origins of certain unique human attributes? (For the purposes of this discussion we define these attributes as including not only walking upright and making use of tools but also an enlarged brain and the ability to communicate by speaking.)

These questions and many more can be answered only after the first question in the series is disposed of. Basically taxonomic in nature, it asks how many species are represented in the East Turkana fossil-hominid assemblage. We have already said that in general two genera were present: *Australopithecus* and *Homo*. How may the two be subdivided?

The answer to this basic and far from simple question is not an easy one. Conspiring against a clear-cut response are such factors as the smallness of the sam-

ple, the fragmentary condition of the individual specimens, the fact that even among individuals of the same species a large degree of morphological variability is far from uncommon and, under this same heading, the fact that a great deal of variation is often found between the two sexes of a single species. Also not to be neglected is the fallibility of the analyst, who is prone to human preconceptions. For example, the very order of discovery of the East Turkana hominids has affected our hypotheses, and we have had to chop and change in order to keep abreast of later discoveries.

It is illuminating in this connection to review the sequence of hominid discoveries in East Turkana. The first specimens to be identified were individuals of the species *Australopithecus robustus*. Fossil representatives of this species were first found at sites in South Africa decades ago. Characteristically they are large of face and massive of jaw; the molar and premolar teeth are very large, although the incisors and canines are small, about the same size as the front teeth of modern man. Although the facial skeleton is large, the brain case is relatively small: the average cranial capacity is about 500 cubic centimeters, compared with the modern human aver-

			ILERET										
	AGE (MILLION YEARS)	TUFF COMPLEX	1/1A	3	5	6/6A	7A	8	10	11	12	15
GUOMDE FORMATION					●3884	✕ 999						
UPPER MEMBER, KOOBI FORA FORMATION	1.22 – 1.32	CHARI										
		MIDDLE/ LOWER	● 725 ✕ 739 ● 728 ✕ 741 ● 805 ✕ 993 ● 3883	● 992 ● 1467 ✕ 740		● 731 ● 1466	● 404			● 726 ✕ 1465		
	1.48 – 1.57		✕ 1463			● 818 ● 2593		● 729 ● 807 ● 733 ● 808 ✖803 ● 806 ● 809		● 1468		
			● 819 ● 820 ● 1817 ● 2595	● 1819		● 727 ● 2592 ● 801 ● 3737 ● 802 ✕ 1464 ● 1170 ✕ 1823 ● 1171 ✕ 1824 ● 1816 ✕ 1825 ● 1818			● 406 ● 407 ● 732 ✕ 815		✕1591 ✕1592	
LOWER MEMBER, KOOBI FORA FORMATION	2.42 (FITCH, MILLER) 1.6 – 1.8 (CURTIS)	(KBS EQUI ALENT)								●1593	● 2597 ● 2599 ● 2598 ✕ 2596
										● 1590		

INVENTORY OF FOSSIL HOMINIDS from East Turkana appears in this chart: specimens, identified by accession numbers, are listed according to the numbered area where they were found. Their positions do not indicate any relative temporal position other than a location between the dated tuff marker layers (or equivalent markers). Such a position means that the specimen is older than the known

age of 1,360 c.c. Because the chewing muscles were evidently of a size commensurate with the large cheek teeth and massive jaws, many *A. robustus* individuals have not only extremely wide-flaring cheekbones but also a bony crest that runs fore and aft along the top of the brain case to provide a greater area for the attachment of chewing muscle.

Specimens of *A. robustus* have also been found in East Africa, most notably by Louis and Mary Leakey at Olduvai Gorge. The East African examples are on the whole even larger than those from South Africa, and their cheek teeth are more massive. A well-known example is "Zinjanthropus," an Olduvai cranium now accepted by most scholars as being closely related to the *A. robustus* specimens from South Africa. (Some scholars, it should be noted, still assign "Zinjanthropus" to a related species of *Australopithecus, A. boisei.*) Such taxonomic niceties aside, the fact is that the East Turkana deposits have been found to contain a good number of fossils that can be placed in this hominid species.

In the early investigations at East Turkana the skulls of certain smaller and less robust hominids were also discovered. Indeed, one such cranium, deformed by crushing, turned up near

(although not in the same stratigraphic horizon as) a robust *Australopithecus* cranium: KNM-ER 406. When this crushed specimen was first discovered, it could not easily be given a taxonomic position. The finding of a second gracile (as opposed to robust) specimen, however, suggested to us that male-female dimorphism might account for both kinds of cranium. In the second specimen most of the right side of the brain case and facial skeleton and part of an upper-jaw premolar tooth and the roots of the molars were preserved. It is evident that although this individual is substantially less robust than KNM-ER 406, its premolar and molar teeth were only a little less massive than those of the robust one. If among the species *A. robustus* the morphological differences between males and females were as great as they are among gorillas, then the robust, crested specimens from East Turkana could be males and the more gracile specimens could be females.

The age of these *Australopithecus* specimens is substantially greater than that of any previously uncovered in East Africa (the age of the South African specimens remains in question), but their discovery presented no taxonomic

problems. This happy state of simplicity came to an end in 1972 with the discovery of the cranium KNM-ER 1470. Bwana Bernard Ngeneo came across it on an exposure of older sediments belonging to the lower member of the Koobi Fora Formation. When he found the specimen, all that could be seen was a scattering of bone fragments on the rock surface. The fragments were relatively fragile, which led us to assume that they had been washed out of the matrix quite recently.

The specimen KNM-ER 1470 is a large, lightly built brain case with a considerable amount of the facial skeleton preserved. Our colleague Ralph L. Holloway, Jr., of Columbia University has determined that its cranial capacity is about 775 c.c. The facial skeleton is very large, and the proportions of the front and cheek teeth are indicated by the preserved tooth sockets and by both the sockets and the broken roots of the molars. The proportions are the reverse of those for *A. robustus;* the incisors and canines are very large and the premolars and molars are only moderately large. Even though the tooth size suggests a formidable chewing apparatus, the brain case shows no sign of a crest for the attachment of heavy chewing

KOOBI FORA

TUFF COMPLEX	118	129	130	131	105	117	116	103	104	119	127	121	123	124
KARARI		− − −												
								✕1807						
OKOTE				● 3230	− − − − −	− − −	− − −	✕ 737	− − − − −					
		● 1805			● 405 ✕738			● 403	● 164 ● 1804		● 1507	● 1506	● 1501 ✕ 1503	
		● 1806			● 1477 ✕1476			● 730	● 810 ● 3733	● 1509	● 1508	✕1809	● 1502 ✕ 1504	● 817
	● 1648				● 1478 ✕3736			● 734	● 811 ✕813		● 1814		● 1811 ✕ 1505	
					● 1479			● 1515	● 812 ✕ 997				● 1813 ✕1810	
					● 1480			● 1820	● 814				● 1821 ✕ 1822	
					● 2607			✖1808	● 816				✖1812	
								✕ 736	● 998					
												PLACEMENT PROVISIONAL	
KBS						− − −							
			● 1462	● 1469 ✕1471			✕3735						
		● 417	● 1800	● 1470 ✕1472	● 3731	● 2602							
			● 2601	● 1474 ✕1473	● 3732	● 2604							
			● 2660	● 1482 ✕1475	● 3734								
			✕1500	● 1801 ✕1481									
				● 1802								SKULL, SKULL FRAGMENTS,	
				● 1803								● JAWS OR TEETH	
				● 1873									
ULU BOR (3.18 MILLION YEARS)	− − −		− − − − −	− − −								✕ OTHER BONES	
						● 2603							
						● 2605						✖ BONES OF BOTH CLASSES	
						● 2606							

age of the tuff layer above it and younger than the known age of the tuff below. Numbers in color identify fossils illustrated on pages 65 through 67. Two geological columns define sedimentary strata; note ages of the tuff marker layers. Broken lines show correlation between Ileret and Koobi Fora tuffs. Placement of specimens listed at the far right is provisional because marker layers are absent there.

muscles. Similarly, the cheekbones, although incomplete, do not suggest the same great width of face that is characteristic of *A. robustus*.

Much has been written about the significance of KNM-ER 1470. We believe that certain hominid specimens found at Olduvai Gorge in broken and fragmentary condition are examples of the same kind of skull. If it is necessary to decide on a taxonomic term for these hominids, the species name may well turn out to be *habilis*. (*Homo habilis* is the name that was given to an early species of the genus *Homo* by Louis Leakey and his colleagues John Napier of the University of London and Phillip V. Tobias of the University of the Witwatersrand. The name was not accepted unanimously by other students of fossil man and has even caused heated argument.)

We ourselves cannot agree on a generic assignment for KNM-ER 1470. One of us (Leakey) prefers to place the species in the genus *Homo*, the other (Walker) in *Australopithecus*. The disagreement is merely one of nomenclature; we are in firm agreement on the evolutionary significance of what are now multiple finds. Since 1972 two additional partial skulls of this large-brained, thin-vaulted kind have been found in association with strata assigned to the lower member of the Koobi Fora Formation.

It was at about the time of the discovery of KNM-ER 1470 that we and our colleagues began to disagree as to the taxonomic position of certain well-preserved lower jaws from the East Turkana region. The initial source of disagreement was a small cranium: KNM-ER 1813. It is the cranium of a small-brained hominid with the average *Australopithecus* cranial capacity: 500 c.c. It has a relatively large facial skeleton and

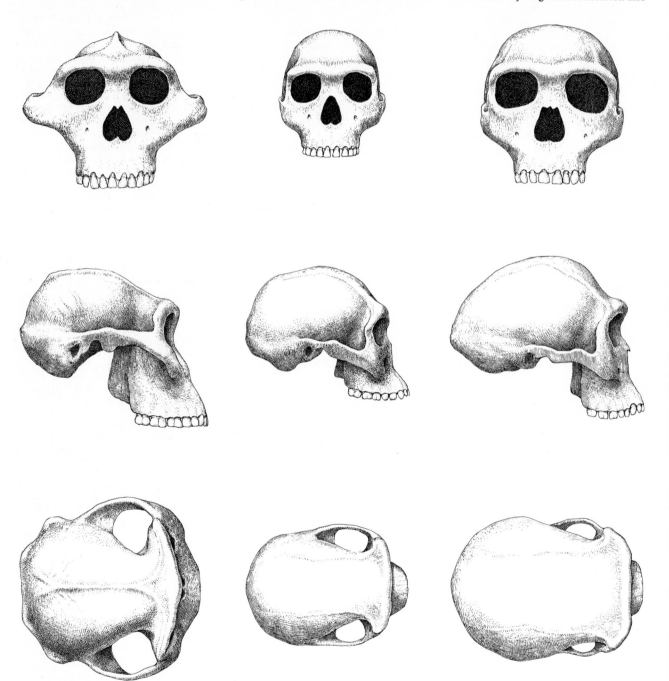

THREE FORMS OF HOMINID are represented among the fossils found in the upper member of the Koobi Fora Formation. The illustration shows them as they would appear if they were restored; lower jaws are omitted. The three may be assigned to particular species in five different ways: all three may belong to a single highly variable species, or they may belong to two species in three possible combinations, or each form may be a valid species on its own. The authors suggest that the three-species hypothesis is the most probable of the five.

SURFACE-SCRAPING PARTY of the Koobi Fora Research Project is seen at work in the Ileret area of East Turkana. The discovery of a hominid jawbone on the surface has led to the collection and screening of loose topsoil and rock in the hope of locating additional fossil fragments. Geologists in the party simultaneously record the stratigraphic position of the exposure and set up a site marker.

FOSSIL JAWBONE is photographed where it was found, exposed on the surface by erosion of the sedimentary rock that once surrounded it. This is the fragmented mandible of a robust member of the genus *Australopithecus;* it was preserved through chance burial in sediments almost two million years ago. Mandibles such as this one are so sturdy that a disproportionate number have survived as fossils.

palate. The upper teeth preserved with the palate are comparatively small, however, and bear a striking resemblance to the teeth of one of the *Homo habilis* specimens from Olduvai, OH-13. It happens that in the initial controversy over the original *H. habilis* specimens, OH-13 represented a species that even skeptics agreed was nearly, if not actually, identical with the species *Homo erectus,* a member in good standing of the genus *Homo* that was first recognized in Java and northern China.

The resemblances between KNM-ER 1813 and OH-13 go further than their teeth. In all the parts that can be compared—the palate as well as the teeth, much of the base of the skull and most of the back of the skull—the two specimens are virtually identical. This leads us to believe the usual reconstructions of OH-13, which have assumed that the specimen had a large cranial capacity and an *erectus*-like skull, are in error. That is not all. The mandible of OH-13 was preserved; its small size and the details of the teeth were major components of the evidence leading to the conclusion that *habilis* was near the *erectus* line. The comparable lower jaws we have found at East Turkana, we can now see, make it clear that the OH-13 mandible could just as well have been hinged to a small-brained, thin-vaulted skull like that of KNM-ER 1813.

To make a final point about these enigmatic East Turkana specimens, we believe there are strong resemblances between them and some of the smaller *Australopithecus* specimens from sites in South Africa, specimens that are usually placed in the gracile species *A. africanus.* Faced with so many possibilities, we argue for caution in the making of taxonomic judgments. Such caution should prevail not only when the evidence in hand is a few isolated teeth but also when the evidence is more generous: lower jaws and upper jaws with the teeth still in place.

In the 1975 season we discovered a remarkably complete cranium: KNM-ER 3733. The find showed unequivocally that a member of our own genus was present in East Turkana when the early strata of the Koobi Fora upper member were formed. The skull bears a striking resemblance to some of the *Homo erectus* skulls found in the 1930's near Peking and is certainly a member of that species. The brain case is large, low and thick-boned. Its principal part is formed by the projecting occipital bone, and its cranial capacity is about 850 c.c. The brow ridges jut out over the eye orbits, and a distinct groove is visible behind them where the frontal bone rises toward the top of the vault.

KNM-ER 3733 has a small face tucked in under the brow ridges. The sockets of some of the upper teeth are preserved. The missing front teeth were relatively large, but the back teeth (some of them still in place) are of only modest proportions. The third molars are among the missing teeth, but the evidence is that they were quite reduced in size. They had been erupted long enough for them to wear grooves in the second molars in front of them, yet the bone forming the sockets indicates that their roots were very small. The point is important because a diminution in the size of this tooth is a common phenomenon in modern human populations.

This fine example of *Homo erectus* from East Turkana predates the example of the same species found at Olduvai Gorge by half a million years and is about a million years older than the examples from northern China. And KNM-ER 3733 is not alone. Another *erectus* specimen was found shortly afterward in Area No. 3 of the Ileret fossil beds. This is KNM-ER 3883. The Ileret specimen is from approximately the same geological horizon as KNM-ER 3733. It has much the same cranial conformation, but its brow ridges, facial skeleton and mastoid processes are somewhat more massive. The cranial capacity of KNM-ER 3883 has not yet been determined, but there is no reason to expect that it will be much different from that of 3733. The similarity of the two East Turkana specimens to specimens from far away that are very much younger strongly suggests that *Homo erectus* was a morphologically stable species of man over a span of at least a million years.

Leaving aside the problems presented by various fragmentary specimens, how can the new hominid finds from East Turkana be assessed taxonomically? One might simply suggest a series of normal taxonomic assignments in the light of what we see at present, acknowledging that as in the past the assignments are likely to be changed. We shall not do so here because we now view the problem of taxonomic assignment in a slightly different way.

Considering only the fossils from the upper member of the Koobi Fora Formation, we think we recognize specimens that might be assigned to three different species. At the same time we may have seriously misunderstood the quantity and quality of variation in any one of the three species. In order to acknowledge this we should like to consider the following possibilities:

1. The three forms are only artifacts of our imagination. Only one hominid species was present, and what we take to be distinct types are merely morphological variants within that species.

This, of course, is the single-species hypothesis. Its strongest proponents have been Loring C. Brace and Milford H. Wolpoff of the University of Michigan. Simply put, the hypothesis states that ever since the human attributes of upright walking, a prolonged childhood,

a large brain and small canine teeth became established there has been only one hominid species. What brought this about, those who favor the hypothesis aver, was culture. Once culture became the human domain—and evidence for this can be sought in the fossil bones by looking for the basic human anatomical attributes that developed along with culture—the human ecological niche became so large that the species with culture always had the edge over any other species in competition for the available resources.

2. Two of the three forms represent one species; the third represents a second species. In this hypothesis the separate species is *Homo erectus*. Both of the other forms, then, must be members of a single, highly variable and sexually dimorphic species. The very robust specimens are males and the very gracile ones are females.

To accept this version of the dimorphic hypothesis one must postulate a great deal of variability. The difference in cranial capacity between the robust and the gracile specimens is within the limits seen in living populations of sexually dimorphic apes. The difference in tooth size, however, is outside the observed limits among living apes.

3. Two of the three forms constitute one species, as above, except that the robust forms represent one of the two species and the *Homo erectus* specimens and the gracile forms together represent another highly variable species.

In this dimorphic hypothesis the principal postulated variation is in the size and shape of the brain case. The admissibility of the hypothesis hinges on accepting the fact that among early humans the cranial capacity of the females was roughly half that of the males.

4. Two of the three forms constitute one species, as above, except that the gracile forms represent one of the two species and the highly variable second species consists of the robust forms and the *Homo erectus* specimens.

In this dimorphic hypothesis the principal postulated variations involve the brain case, the jaws and the teeth. The admissibility of the hypothesis hinges on accepting the fact that small-brained but large-jawed forms and large-brained but small-jawed forms can be placed together in the same species.

5. The three forms represent three separate species.

Having listed the five possible hypotheses, we can now assess the probability of each being correct. We shall do so bearing in mind both the fossil evidence and what is known about the variability of living primate populations. We think the probability that the single-species hypothesis is correct is very low. First, the hypothesis involves accepting the fact that there is an enormous amount of intraspecies variability. Second, we think, along with others, that the adaptations apparent in the skulls of both extreme forms (*Homo erectus* and *Australopithecus robustus*) are different. In *H. erectus* the size of the brain case seems to overwhelm the chewing apparatus, as it does in living man. In *A. robustus* the opposite is true.

For the same reasons we think the

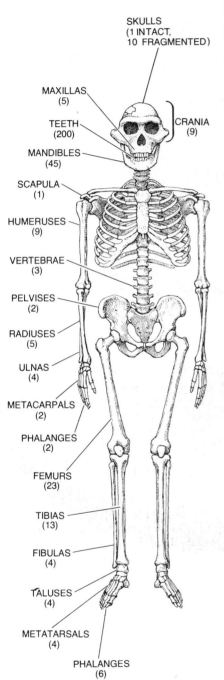

SKULLS
(1 INTACT,
10 FRAGMENTED)

MAXILLAS
(5)

CRANIA
(9)

TEETH
(200)

MANDIBLES
(45)

SCAPULA
(1)

HUMERUSES
(9)

VERTEBRAE
(3)

PELVISES
(2)

RADIUSES
(5)

ULNAS
(4)

METACARPALS
(2)

PHALANGES
(2)

FEMURS
(23)

TIBIAS
(13)

FIBULAS
(4)

TALUSES
(4)

METATARSALS
(4)

PHALANGES
(6)

DISPROPORTION in preservation of the remains of fossil hominids unearthed thus far in East Turkana is assessed in this illustration. Teeth, which are the hardest of all body parts, are the most numerous remains. Mandibles, most of them representing the robust species of *Australopithecus*, come next. Rarest of all are pelvic bones, the bones of hands and feet, vertebrae and the bones of the lower arms.

probability that the fourth hypothesis is correct is very low. In addition difficulties other than anatomical ones stand in the way. One must also ask why no specimens of *A. robustus* are found in Java and China, where *H. erectus* specimens are comparatively abundant. If the answer is that such specimens were present but for some reason were not fossilized or have not yet been unearthed, then how is it that *A. robustus* is the commonest hominid in the East African fossil record?

The third hypothesis is an attractive one, but we think the probability of its being correct is low. It would be difficult enough to account for the enormous dimorphism in brain size without having to supply an answer to an additional question: Why are these gracile forms of *Australopithecus* not found as fossils in Java and China?

The second hypothesis, we think, is more likely to be correct than the third. Although the variability in the dimensions and proportions of the teeth cannot be matched in living dimorphic populations, there is some hint that dental dimorphism might have been greater in extinct hominoids: the superfamily that includes both the hominids and the apes. Louis de Bonis of the University of Paris has collected a series of Miocene hominoid jawbones in Macedonia. His sample is from one small area; if the mandibles all belong to the same species, they represent a degree of dental dimorphism greater than that found in living anthropoid apes.

The hypothesis with the best chance of being correct, we believe, is the last of the five: that three species are present. Sorted out in this way, none of the specimens within each group shows more variability in brain size and chewing apparatus than we see among living anthropoids. Although the fossil record from the lower member of the Koobi Fora Formation is far less rich than that from the upper member, a similar three-species hypothesis could also be advanced with respect to the specimens found in it. In this hypothesis the third species in addition to the robust and gracile ones would be represented by the *habilis* specimens.

Several consequences follow from our probability assessments. For example, in our view the demonstration at East Turkana that *Homo erectus* was contemporaneous with some of the largest representatives of *Australopithecus robustus* amounts to a disproof of the single-species hypothesis. We believe both *H. erectus* and *A. robustus* had essentially human characteristics. If this is the case, it follows that they occupied separate ecological niches. It would seem either that one of the species did not possess culture and yet still developed human characteristics or that the argument that the advantage of culture

would give the cultured hominid dominance within a very wide ecological niche is flawed.

We prefer the first alternative, and we would nominate *Australopithecus* for the role of the hominid without culture. Accepting this alternative requires that we keep searching for the natural-selection pressures that have been responsible for producing the basic human attributes.

Where did *Homo erectus* come from? Some have suggested that the species arose in Asia and migrated to Africa. This seems to us an unnecessarily complicated hypothesis. For one thing, it neglects *habilis*. Worse, it implies that a population of these large-brained hominids, who presumably made the stone tools found in the early East Turkana strata, evolved independently in Africa at the right time to fit into an ancestor-descendant relation with *Homo erectus* and then came to an abrupt halt, without playing any further part in human evolution.

It is our view that the *habilis* populations are directly antecedent to *Homo erectus*. If the earlier range of dates for the strata where the *habilis* specimen KNM-ER 1470 was found proves to be correct, then the transition from *habilis* to *erectus* could have been a gradual one, spanning a period of well over a million years. If the later dates are correct, then the transition must have been very rapid indeed.

W e have concentrated here on giving an overview of the hominid record in East Turkana to the neglect of other work in progress that promises to answer some of the questions we have raised. The Koobi Fora Research Project is continuing its field activities, and a number of special studies are also under way outside Africa. Michael H. Day of St. Thomas's Hospital Medical School in London is examining the fossil limb bones and associated parts from East Turkana to see what can be learned about the various species' capacity for upright walking. Bernard A. Wood of the Middlesex Hospital Medical School in London is conducting a full analysis of the skulls and jaws in order to document the extremes and means of variability and dimorphism. Holloway is studying casts of the inside of brain cases in an effort to trace the evolution of the brain. One of us (Leakey) continues as codirector of the research project and the other (Walker) is studying the biomechanics of hominid mastication, examining the fossil teeth from East Turkana with the scanning electron microscope in an attempt to deduce dietary habits from the patterns of tooth wear. These studies and others aim at reconstructing as much as can be reconstructed about the biology of these very early hominids in the hope of determining just what it was that made us human.

7

The Casts of Fossil Hominid Brains

by Ralph L. Holloway
July 1974

The skulls of man and his precursors can be used as molds to make replicas of the brain. These casts indicate that man's brain began to differ from that of other primates some three million years ago

Man is not the largest of the primates (the gorilla is larger), but he has the largest brain. How did this come about? The question is hardly a new one, but a considerable amount of new evidence is now available to those in search of the answer. In brief the evidence suggests that, contrary to what is widely believed, the human brain was not among the last human organs to evolve but among the first. Neurologically speaking, brains whose organization was essentially human were already in existence some three million years ago.

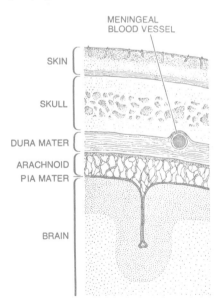

MENINGEAL
BLOOD VESSEL

SKIN

SKULL

DURA MATER

ARACHNOID
PIA MATER

BRAIN

LAYERS OF TISSUE lie between the surface of the brain and the cranium. They include the pia mater, the arachnoid tissue that contains the cerebrospinal fluid, and the thick dura mater. Meningeal blood vessels (*color*) often leave a clear imprint on the inner surface of the cranium. The sutures of the cranial bones may also show, but among higher primates the convolutions of the cerebral cortex are not often apparent.

The brain is the most complex organ in the primate body or, if one prefers, the most complex set of interacting organs. It consists of a very large number of interconnected nuclei and fiber systems, and its cells number in the billions. With certain exceptions, however, the cortex, or upper layer, of one primate brain exhibits much the same gross morphology as the cortex of another, regardless of the animals' relative taxonomic status. Whether the animal is a prosimian (for example a lemur), a monkey, a pongid (an ape) or a hominid (a member of the family that includes the genus *Homo*), what varies from one primate brain to another is not so much the appearance of the cortex as the fraction of its total area that is devoted to each of the major cortical subdivisions. For instance, whereas the chimpanzee is taxonomically man's closest living primate relative and the brains of both look much alike superficially, the chimpanzee brain is significantly different from the human brain in the relative size and shape of its frontal, temporal, parietal and occipital lobes. The differences are reflected in the different position of the sulci, or furrows, that mark the boundaries between lobes, and in the differently shaped gyri, or convolutions, of the lobes themselves.

The comparative neurology of living primates can of course cast only the most indirect light on human brain evolution. After all, the living representatives of each primate line have reached their own separate evolutionary pinnacle; their brains do not "recapitulate" the evolutionary pathway that man has followed. The neurological evidence is nonetheless invaluable in another respect. It is the only source of information linking aspects of gross brain morphology with aspects of behavior. The results of a great many primate studies allow some valid generalizations on this subject. For example, it is well established that the occipital lobe of the cortex, at the back of the brain, is involved in vision. The parietal lobe is involved in sensory integration and association, the frontal lobe in motor behavior and the more complex aspects of adaptive behavior, and the temporal lobe in memory. It is the interactions among these gross cortical divisions, and also among the subcortical nuclei and fiber tracts, that organize coordinated behavior.

It is also a valid generalization to say that a gross brain morphology which emphasizes relatively small temporal and parietal lobes and a relatively large area of occipital cortex is neurologically organized in the pongid mode and is thus representative of the apes' line of evolutionary advance. Conversely, a gross morphology that emphasizes a reduced area of occipital cortex, particularly toward the sides of the brain, and an enlarged parietal and temporal cortex is hominid in its neurological organization. It follows that any evidence on the neurological organization of early primates, including hominids and putative hominids, is of much importance in tracing the evolution of the human brain.

Such evidence is of three kinds: direct, indirect and inferential. The only direct evidence comes from the study of endocranial casts, that is, either a chance impression of a skull interior that is preserved in fossil form or a contemporary man-made replica of the interior of a fossil skull. The indirect evidence is of two kinds. The first is a by-product of endocranial studies; it consists of the conclusions that can be drawn from a comparison of brain sizes. To draw conclusions of this kind, however, can involve some degree of acceptance of the

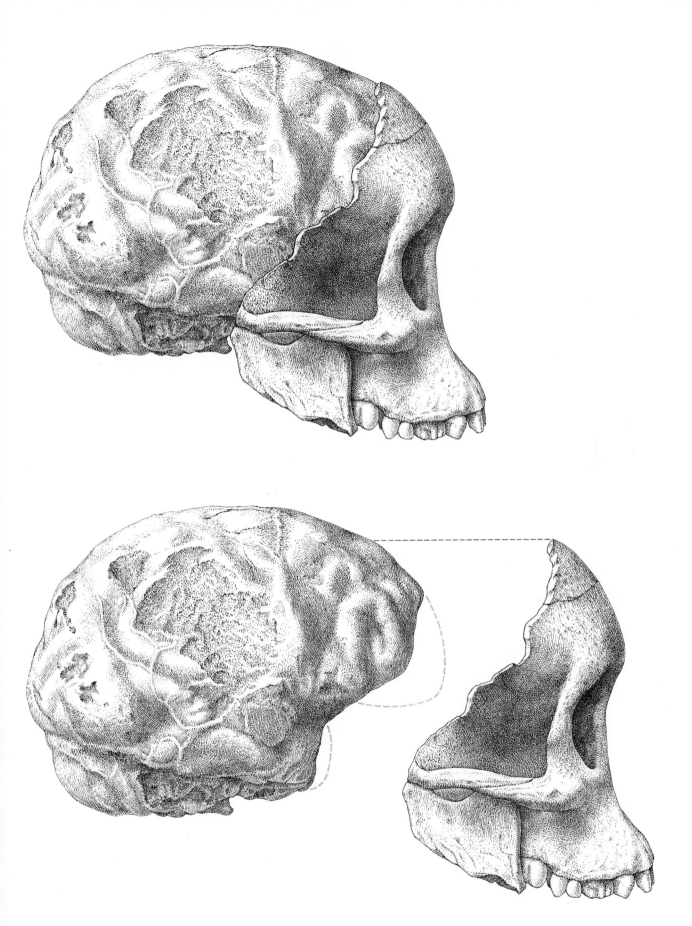

TAUNG JUVENILE, the first specimen of *Australopithecus* to be unearthed, is shown in the top drawing with a portion of the fossilized skull (including the facial bones, the upper jaw and a part of the lower jaw) in place on the natural cast of its brain. The cast is seen separately in the bottom drawing; parts of the frontal and temporal lobes that were not preserved are indicated. The estimated brain volume of an adult of this lineage, once calculated at about 525 cubic centimeters, has now been scaled down to 440 c.c.

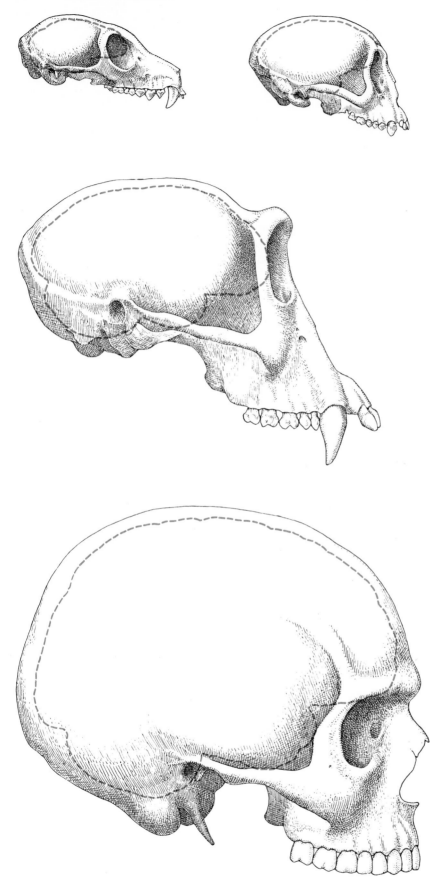

INCREASE IN BRAIN SIZE among primates is apparent in the different dimensions of the skull of a prosimian (*top left*), a New World monkey (*top right*), a great ape (*second from bottom*) and modern man (*bottom*). The brains (*color*) of the latter three are illustrated on the opposite page; skulls are reproduced at approximately half actual size.

questionable premise that there is a correlation between brain size and behavioral capacity. The second kind of indirect evidence comes from the study of other fossil remains: primate hand and foot bones, limb bones, pelvises, vertebral components, jaws and teeth. From these noncranial parts conclusions can be drawn about body size and behavioral capabilities such as bipedal locomotion, upright posture, manual dexterity and even mastication. Various patterns of musculoskeletal organization, of course, reflect matching variations in neurological organization.

Finally there is inferential evidence, particularly with respect to early hominid evolution. This can be described as fossilized behavior. The inferential evidence includes stone tools that exhibit various degrees of standardization (suggesting "cultural norms") and bone debris that reveals what animals the hominids selected as their prey. Any evidence of activity provides some grounds for inference about the general state of the individuals' neurological organization. Our concern here, however, will be with the direct endocranial evidence and with a part of the indirect evidence.

The convolutions of the cerebral cortex and its boundary furrows are kept from leaving precise impressions on the internal bony table of the skull by the brain's surrounding bath of cerebrospinal fluid and by such protective tissues as the pia mater, the arachnoid tissue and the dura mater, or outer envelope. In mammals the extent of masking varies from species to species and with the size and age of the individual. The skull interiors of apes and men, both living and fossil, are notable for bearing only a minimal impression of the brain surface. In virtually all cases the only detailed features that can be traced on the endocranial cast of a higher primate are the paths of the meningeal blood vessels. Depending on the fossil's state of preservation, however, even a relatively featureless cast will reveal at least the general proportions and shape of the brain. This kind of information can be indicative of a pongid neurological organization or a hominid one.

Just how far back in time can the distinction between pongid and hominid brains be pursued? There is a barrier represented by the key Miocene primate fossil *Ramapithecus*: no skull of the animal has been discovered. *Ramapithecus* flourished some 12 million to 15 million years ago both in Africa and in Asia. Elwyn L. Simons and David Pilbeam of Yale University have proposed that it is a hominid, but its only known remains are teeth and fragments of jaws. For the present we must make do with more numerous, although still rare, primate skulls that are many millions of years more recent.

The first of these skulls became known in 1925, when Raymond A. Dart of the University of Witwatersrand described the fossilized remains of a subadult primate that had been discovered in a limestone quarry at Taung in South Africa. The find consisted of a broken lower jaw, an upper jaw, facial bones, a partial cranium and a natural endocranial cast [*see illustrations on page 75*]. In the half-century since Dart named the specimen *Australopithecus,* or "southern ape," seven other skulls of *Australopithecus* (six of them internally measurable) have been found in South Africa and from three to six more in East Africa.

"Three to six" refers not to any uncertainty about how many skulls have been found but to how they are to be assigned to one or another genus or species of hominid. For example, the genus *Australopithecus* consists of two species: *A. africanus,* the "gracile," or lightly built, form to which the Taung fossils belong, and *A. robustus,* a larger, more heavily built species [*see illustration on page 81*]. Of the eight South African skulls, six are gracile, one is robust and the eighth (a specimen from Makapansgat designated MLD 1) is not definitely assigned to either species. In the same way three of the East African skulls are unanimously assigned to the robust species of *Australopithecus.* The other three, designated *Homo habilis* by their discoverers, Louis and Mary Leakey, are classed with the gracile species by some students of the subject, but they are still generally known by the name the Leakeys gave them. As will become apparent, there are grounds for preserving the distinction.

Fortunately, no matter what controversy may surround the question of how these early African hominids are related to one another, it has very little bearing on the question of their neurological development. The reason is that in each instance where an endocast is available, whether the skull is less than a million years old or more than two million years old, the brain shows the distinctive pattern of hominid neurological organization. Let us review the cortical landmarks that distinguish between the pongids and the hominids. Starting with the frontal lobe, the *Australopithecus* endocasts show a more hominid pattern in the third inferior frontal gyrus, being larger and more convoluted than endo-

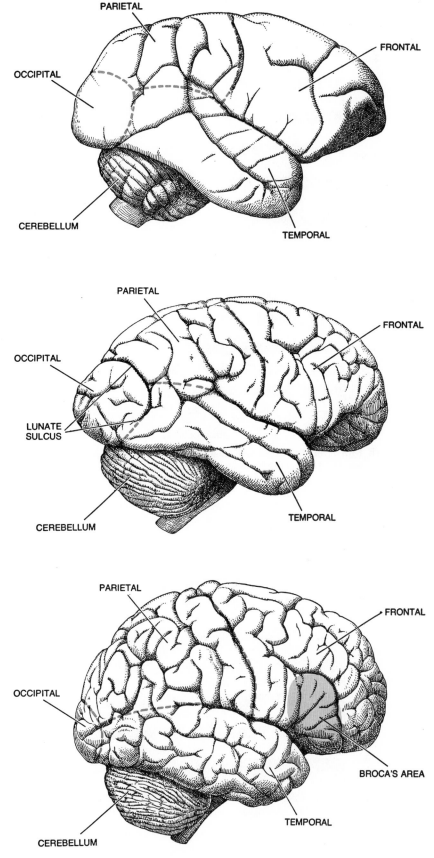

GROSS DIFFERENCES in the neurological organization of three primate brains are apparent in the size of the cerebral components of a ceboid monkey (*top*), a chimpanzee (*middle*) and modern man (*bottom*). The small occipital lobe and the large parietal and temporal lobes in man, compared with the other primates, typify the hominid pattern. Lunate sulcus, or furrow (*color*), on the chimpanzee's brain bounds its large occipital lobe.

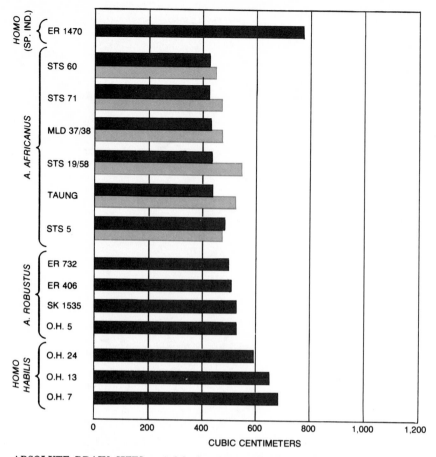

ABSOLUTE BRAIN SIZES varied both among and between the four earliest African hominid species. The 14 specimens are shown here in order of increasing brain volume, except for the earliest of all: the three-million-year-old specimen from East Rudolf, ER 1470, at the head of the list. The other abbreviations stand for Sterkfontein (STS), Makapansgat (MLD), Olduvai hominid (O.H.) and Swartkrans (SK). Double bars show former (*color*) and present calculated brain sizes of six specimens; the Taung volumes are adult values.

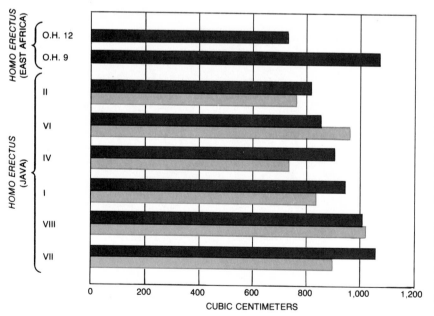

LATER HOMINID'S BRAINS were larger both on the average and absolutely, with the exception of one doubtful specimen from Olduvai Gorge. Shown here are the brain volumes for the seven certain specimens and one doubtful specimen of *Homo erectus*; two are from East Africa and six are from Java. Double bars show former (*color*) and present calculated brain sizes of the Java specimens. Casts of other *H. erectus* brains are not available.

casts of pongid brains of the same size. The orbital surface of their frontal lobes displays a typically human morphology rather than being marked by the slender, forward-pointing olfactory rostrum of the pongid. The height of the brain, from the anterior tips of the temporal lobes to the summit of the cerebral cortex, is proportionately greater than it is in pongids, suggesting an expansion of the parietal and temporal lobes and a more "flexed" cranium. Moreover, the temporal lobes and particularly their anterior tips show the hominid configuration and not the pongid one.

Another landmark of neurological organization provides further evidence, in part negative. This is the lunate sulcus, the furrow that defines the boundary between the occipital cortex and the adjacent parietal cortex. In all ape brains the lunate sulcus lies relatively far forward on the ascending curve of the back of the brain. The position is indicative of an enlarged occipital lobe. In modern man, when the lunate sulcus appears at all (which is in fewer than 10 percent of cases), it lies much closer to the far end of the occipital pole. On those *Australopithecus* endocasts where the feature can be located, the lunate sulcus is found in the human position, which indicates that the (perhaps associative) parietal lobes of the *Australopithecus* brain were enlarged far beyond what is the pongid norm.

It may seem surprising that *Australopithecus*, a genus that has only in recent years been granted hominid status on other anatomical grounds, should have an essentially human brain. It has certainly been a surprise to those who view the human brain as a comparatively recent product of evolution. The finding is not, however, the only surprise of this kind. The most remarkable new fossil primate discovery in Africa is the skull known formally as ER 1470, found by Richard Leakey and his colleagues in the region east of Lake Rudolf in Kenya in 1972. The fossil is nearly three million years old.

Through the courtesy of its discoverer I recently made an endocranial cast of ER 1470. Two facts were immediately apparent. Not only had the skull contained a brain substantially larger than the brain of either the gracile or the robust species of *Australopithecus* (and that of *Homo habilis* too) but also this very ancient and relatively large brain was essentially human in neurological organization. Leakey's find pushes the history of hominid brain evolution back in time at least as far as the shadowy

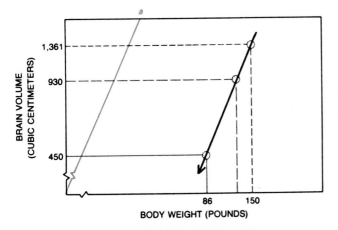

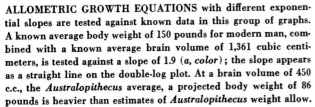

ALLOMETRIC GROWTH EQUATIONS with different exponential slopes are tested against known data in this group of graphs. A known average body weight of 150 pounds for modern man, combined with a known average brain volume of 1,361 cubic centimeters, is tested against a slope of 1.9 (*a, color*); the slope appears as a straight line on the double-log plot. At a brain volume of 450 c.c., the *Australopithecus* average, a projected body weight of 86 pounds is heavier than estimates of *Australopithecus* weight allow.

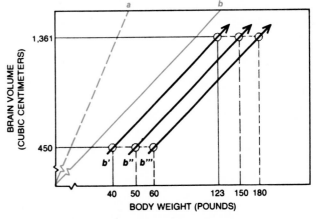

LESS RADICAL SLOPE, 1.0 (*b, color*), is tested in this graph against three different estimates of the average body weight of *Australopithecus*: 40 pounds, 50 pounds and 60 pounds. Where the parallel exponential slopes intersect the average cranial capacity for modern man (*b', b'', b'''*) only the 50-pound body weight estimated for *Australopithecus* yields a projected body weight for man that agrees with the known average. An *Australopithecus* body weight much above or below 50 pounds thus appears improbable.

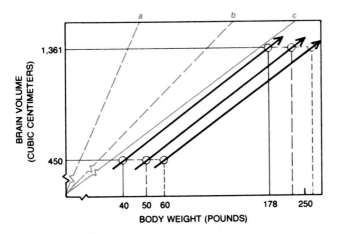

STILL A THIRD SLOPE, .66 (*c, color*), which is usually the most favorable allometric rate for mammals, is tested against the same three *Australopithecus* body-weight estimates. The .66 slope proves clearly unsuited to hominids. When even the minimum weight for *Australopithecus* is projected, the predicted weight for modern man is excessive; added weight yields even more grotesque results.

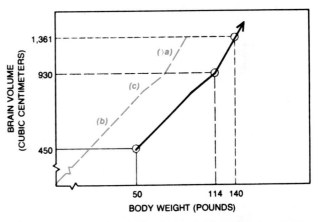

COMBINATION OF SLOPES tests the assumption that hominid rates of growth differed at different times. The zigzag predicts a plausible 114-pound body weight for *Homo erectus* with an average-size brain (930 c.c.) and a body weight for modern man not far below average. The implication is that rates of growth probably did vary, thereby varying selection pressures for changes in brain size.

boundary between the Pliocene and Pleistocene eras.

To summarize the direct evidence of the endocasts, it is now clear that primates with essentially human brains existed some three million years ago. Let us go on to see what details, if any, the indirect evidence of brain size can add to this finding. Here it is necessary, however, to state an important qualification. So far as modern man is concerned, at least, no discernible association between brain size and behavior can be demonstrated. On the average, to be sure, *Homo sapiens* is a big-brained primate. What is sometimes forgotten, however, is that the human average embraces

some remarkable extremes. The "normal" sapient range is from 1,200 cubic centimeters to 1,800, but nonpathological brains that measure below 1,000 c.c. and above 2,000 are not uncommon. Moreover, this range of more than 1,000 c.c. is in itself greater than the average difference in brain size between *Australopithecus* and *H. sapiens*.

As to how such variations may affect behavior, it is probably sufficient to note that the brains of Jonathan Swift and Ivan Turgenev exceeded 2,000 c.c. in volume, whereas Anatole France made do with 1,000 c.c. Clearly the size of the human brain is of less importance than its neurological organization. This con-

clusion is supported by studies of pathology. For example, microcephaly is a disease characterized by the association of a body of normal size with an abnormally small brain. The brain may have a volume of 600 c.c., which is smaller than some gorilla brains. (The gorilla average is 498 c.c., and brains almost 200 c.c. larger have been measured.) Humans with microcephaly are quite subnormal in intelligence, but they still show specifically human behavioral patterns, including the capacity to learn language symbols and to utilize them.

I have recently made complete or partial endocasts of 15 early fossil hominids from South and East Africa. These

SQUIRREL MONKEY	1:12
PORPOISE	1:38
HOUSE MOUSE	1:40
TREE SHREW	1:40
MODERN MAN	1:45
MACAQUE	1:170
GORILLA	1:200
ELEPHANT	1:600
BLUE WHALE	1:10,000

WEIGHT OF THE BRAIN, expressed as a proportion of the total body weight, varies widely among mammals. In spite of his large brain modern man is far from showing the highest proportional brain weight. The proportional weight for man, however, is greater than that of other higher primates. So too, it appears, was the proportional weight for the hominid predecessors of modern man.

endocasts, together with natural ones, have allowed me to calculate the specimens' brain size [see top illustration on page 78]. Leaving aside for the moment two specimens from Olduvai Gorge (one of them certainly *Homo erectus* and the other possibly so), my findings are as follows. First, the brains of most of the South African specimens were substantially smaller than had previously been calculated. Of the six gracile specimens of *Australopithecus* from South Africa none had brains that exceeded 500 c.c. in volume, and most were well below 450 c.c. In contrast, none of the four specimens of the robust *Australopithecus* species had a brain volume of less than 500 c.c. Moreover, two of the four, the one from South Africa and one from East Africa, had brains measuring 530 c.c. Of the three representatives of *Homo habilis*, all from Olduvai, the one with the smallest brain was hominid No. 24, with a possibly overestimated cranial capacity of 590 c.c. The brains of the other two respectively measured 650 c.c. and 687 c.c.

Now, one reason—perhaps the main reason—students of human evolution were so slow to accept *Australopithecus africanus* as a hominid, even though Dart had emphasized the manlike appearance of the Taung endocast from the first, was that its estimated cranial capacities were so small. The brains of some apes, particularly the brains of gorillas, were known to be larger. In fact, as we now know, the brains of the gracile species of *Australopithecus* are even smaller than was originally thought. Since it is now also evident that the *Australopithecus* brains were essentially

human in neurological organization, what are we to make of their surprisingly small size? The answer, it seems to me, is that in all likelihood the size of *Australopithecus'* brain bore the same proportional relation to the size of its body that modern man's brain does to his body.

This contention cannot be proved beyond doubt on the basis of the *Australopithecus* fossils known today. One can estimate body weight on the basis of known height, but the height of both the gracile and the robust species must also be estimated, on the basis of an imperfect sample of limb bones. Even though guesswork is involved, however, it is still possible to estimate various body weights and relate each of these guesses to the species' known brain size. One can then see how the results compare with the known brain-to-body ratio among the other mammals, particularly the primates.

The exponential relation between variations in overall body size and in the size of a specific organ, known technically as allometry, has been studied for nearly a century, so that the brain-to-body ratio is well known for a number of animals [see illustration on this page]. The brain-to-body ratio of modern man is not first among mammals; various marine mammals, led by the porpoise *Tursiops*, rank higher than man. So does one small primate, the ceboid squirrel monkey. However, among the hominoids, that is, the subdivision of primates that includes both the apes and man, modern man's brain-to-body ratio does rank first. Depending to some extent on what weight estimate one accepts for *Australopithecus*, this early hominid appears to have enjoyed a position similar to modern man's.

Thanks to Heinz Stephan and his colleagues at the Max Planck Institute for Brain Research in Frankfurt, an assessment of the relation between brain size and body size has become available that is subtler than simple allometry. Stephan's group has been collecting quantitative data with respect to animal brains for many years. The size and weight of various parts of the brain are measured, and these measurements are related to similar measurements of the animals' entire brain and body. One outcome of the work has been the development of what Stephan calls a progression index. It is the ratio between an animal's brain weight and what the brain weight would have been if the animal had belonged to another species with the same body dimensions. As their standard in this com-

parison Stephan and his colleagues use the brain-to-body ratio of "basal insectivores" such as the tree shrew, which are representative of the original stock from which all the primates evolved.

When the progression index is calculated for modern man, assuming an average body weight of 150 pounds and an average cranial capacity of 1,361 c.c., the resulting index value is 28.8. If different body weights are used with the same average cranial capacity and vice versa, the range of progression indexes for modern man extends from a minimum of 19.0 to a maximum of 53.0. It is interesting to compare the range of the human progression indexes with the indexes of other primates, including those based on estimated brain-to-body ratios for the two species of *Australopithecus*. For example, the maximum progression index for chimpanzees is 12.0. If one works the formula backward and assumes a chimpanzeelike progression index for the gracile species of *Australopithecus*, using an average cranial capacity of 442 c.c., the required body weight turns out to be about 100 pounds. That is nearly twice the maximum estimated weight for the species.

If one uses the same average cranial capacity and assumes that the body weight of the gracile species was only 40 pounds, the progression index is 21.4, which is well within the human range and comfortably close to the human average. When the body weight of the gracile species is estimated at 50 and 60 pounds and the body weight of the robust species is estimated at 60 and 75 pounds, the resulting progression indexes also fall close to or within the human range.

To recapitulate, both the direct evidence of neurological organization and the indirect evidence of comparative brain size appear to indicate that *Australopithecus* and at least one other African primate of the period from three million to one million years ago had brains that were essentially human in organization and that *Australopithecus* was also probably within the human range of sizes with respect to the proportion of the brain to the body. That the brain was small in absolute size, particularly in the gracile species of *Australopithecus*, therefore seems to be without significance. The ratio of brain size to body was appropriate. So far as the subsequent absolute cranial enlargement is concerned, the major mechanism involved, although surely not the only one, appears to have been that as hominids grew larger in body their brains enlarged proportionately. Quite possibly this expansion progressed at different rates at different times [see illustrations on page 79]. In any event, for at least the past three million years there has been no kind of cranial Rubicon waiting to be crossed.

Considering the gaps in the primate fossil record, it is remarkable that so many specimens of a hominid intermediate between the earliest African fossil forms and modern man have been unearthed. The intermediate form is *Homo erectus*. The skulls and fragmentary postcranial remains of that species have been found at various sites in China and in Java, and more fossils are steadily being turned up in both areas. There is also one specimen (and possibly a second) from Olduvai and, depending on one's choice of authorities, one from southern Africa, one from eastern Europe and perhaps two from northern Africa. With the possible exception of the Olduvai and Java fossils, no *H. erectus* specimen is much more than 500,000 years old, and one or more may be a great deal younger.

I have made endocasts of five of the six available *H. erectus* specimens from Java and of both the certain Olduvai specimen and the doubtful one. The *erectus* fossil with the largest cranial capacity is Olduvai hominid No. 9; its brain measures 1,067 c.c. Omitting the doubtful Olduvai specimen, the *erectus* fossil with the smallest brain is one of those from Java. Its cranial capacity is 815 c.c., only 40 c.c. larger than the capacity of the three-million-year-old East Rudolf skull. The average *erectus* cranial capacity, based on the endocasts I have made, is 930 c.c.

Working with this average brain size and estimating the body weight of *H. erectus* to have been 92 pounds, one finds that the Stephan progression index for the species is 26.6. That is remark-

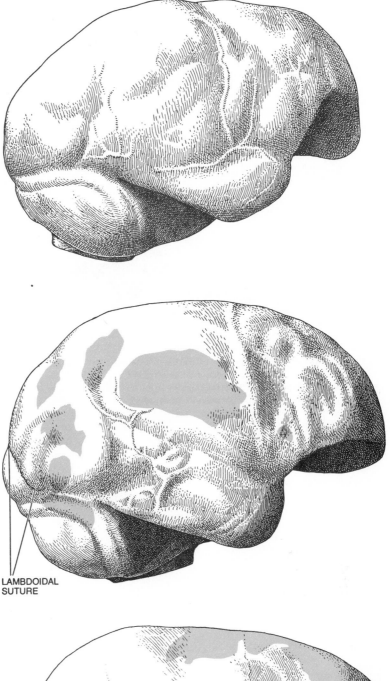

LAMBDOIDAL SUTURE

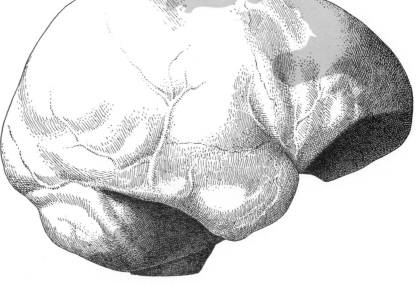

ENDOCRANIAL CASTS are those of (*top*) a chimpanzee, *Pan troglodytes*, (*middle*) a gracile *Australopithecus africanus*, and (*bottom*) the other species, *A. robustus* (color indicates restored areas). In all three casts details of the gyral and sulcal markings of the cerebral cortex are minimal. A differing neurological organization, however, ·can be seen. Both of the hominid brains are higher, particularly in the parietal region. Orbital surface of their frontal lobes is displaced downward in contrast to chimpanzee's forward-thrusting olfactory rostrum. The location of the hominids' lunate sulcus, indicated (*middle*) by suture markings, implies a far smaller occipital lobe than the ape's.

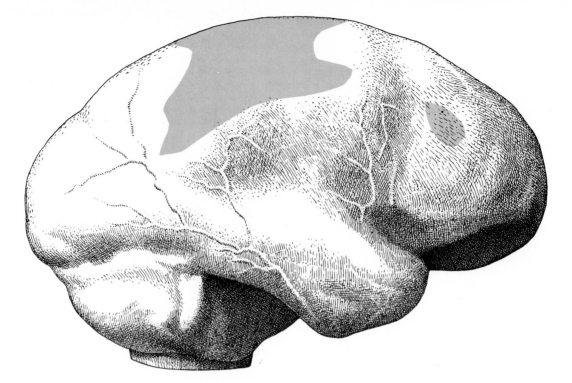

BRAIN CAST OF HOMO ERECTUS shows similar evidence of human neurological organization. As is true of most human cranial casts, the position of the lunate sulcus cannot be determined but the expansion of the temporal lobe and the human shape of the frontal lobe are evident. This is a cast of Java specimen VIII (1969); it reflects the flat-topped skull conformation typical of the fossil forms of *H. erectus* found in Indonesia. Like endocranial casts on preceding page and below it is shown 90 percent actual size.

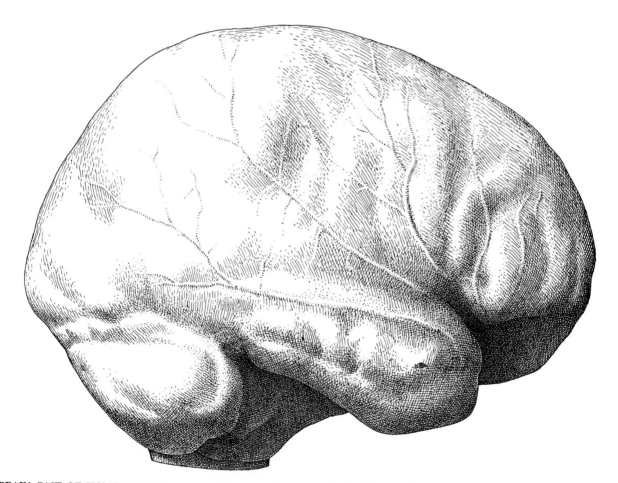

BRAIN CAST OF HOMO SAPIENS was made from a cranium in the collection at Columbia University. The height of the cerebral cortex, measured from its summit to the tip of the temporal lobe, and the fully rounded, expanded frontal lobe, showing a strong development of Broca's area (*see illustration on page* 77), typify the characteristic *H. sapiens* pattern of neurological organization.

ably close to modern man's average of 28.8. Even if the body-weight estimate is raised by some 30 pounds, the progression index falls only to 22.0. This being the case, it is difficult to escape the conclusion that, like *Australopithecus*, *H. erectus* possessed a brain that had become enlarged in proportion to the enlargement of the body. Although the average *erectus* brain is smaller than the average brain of modern man, it nonetheless conforms in gross morphology to the species' status as a recognized member of the genus *Homo* [*see top illustration on opposite page*].

I am not suggesting that any simple, straight-line progression connects the earliest African hominids, by way of *Homo erectus*, to modern man. As others have noted, it is the investigator's mind, and not the evidence, that tends to follow straight lines. The data are still far too scanty to trace the detailed progress of the human brain during the course of hominid evolution. For example, it would not even be safe to assume that, after attaining the stage represented by the earliest-known hominid endocasts, the brain thereafter consistently increased in size. Perhaps there were times when on the average brain sizes decreased simply because body sizes also decreased. One might even go so far as to speculate, and it would be pure speculation, that we are seeing something like this when we observe that the specimens of *Homo habilis* at Olduvai had substantially smaller brains than the far older East Rudolf hominid did.

Some generalizations are nonetheless possible. First, both the direct evidence of the endocasts and the indirect evidence of comparative cranial capacities indicate that the human brain appeared very much earlier than the time when *H. erectus* emerged, perhaps 500,000 years ago. Second, it can be inferred that the emergence of the human brain was paralleled by the initiation of human social behavior. It is not appropriate here to review the evidence of relations between nutrition and behavioral development or of those between the endocrine system and brain growth. Still, it is obvious that brains do not operate in a vacuum and that a part of the nourishment the brain requires is social as well as dietary. Much of the humanness of man's brain is the result of social evolution. The weight of the inferential evidence today suggests that the genesis has been long in the making. It may well predate such elements of fossil behavior as the systematic use of stone tools and the large-scale practice of hunting.

8   Homo Erectus

by William W. Howells
November 1966

This species, until recently known by a multiplicity of other names, was probably the immediate predecessor of modern man. It now seems possible that the transition took place some 500,000 years ago

In 1891 Eugène Dubois, a young Dutch anatomist bent on discovering early man, was examining a fossil-rich layer of gravels beside the Solo River in Java. He found what he was after: an ancient human skull. The next year he discovered in the same formation a human thighbone. These two fossils, now known to be more than 700,000 years old, were the first remains to be found of the prehistoric human species known today as *Homo erectus*. It is appropriate on the 75th anniversary of Dubois's discovery to review how our understanding of this early man has been broadened and clarified by more recent discoveries of fossil men of similar antiquity and the same general characteristics, so that *Homo erectus* is now viewed as representing a major stage in the evolution of man. Also of interest, although of less consequence, is the way in which the name *Homo erectus*, now accepted by many scholars, has been chosen after a long period during which "scientific" names for human fossils were bestowed rather capriciously.

Man first received his formal name in 1758, when Carolus Linnaeus called him *Homo sapiens*. Linnaeus was trying simply to bring order to the world of living things by distinguishing each species of plant and animal from every other and by arranging them all in a hierarchical system. Considering living men, he recognized them quite correctly as one species in the system. The two centuries that followed Linnaeus saw first the establishment of evolutionary theory and then the realization of its genetic foundations; as a result ideas on the relations of species as units of plant and animal life have become considerably more complex. For example, a species can form two or more new species, which Linnaeus originally thought was impossible. By today's definition a spe-

cies typically consists of a series of local or regional populations that may exhibit minor differences of form or color but that otherwise share a common genetic structure and pool of genes and are thus able to interbreed across population lines. Only when two such populations have gradually undergone so many different changes in their genetic makeup that the likelihood of their interbreeding falls below a critical point are they genetically cut off from each other and do they become separate species. Alternatively, over a great many generations an equivalent amount of change will take place in the same population, so that its later form will be recognized as a species different from the earlier. This kind of difference, of course, cannot be put to the test of interbreeding and can only be judged by the physical form of the fossils involved.

In the case of living man there is no reason to revise Linnaeus' assignment: *Homo sapiens* is a good, typical species. Evolution, however, was not in Linnaeus' ken. He never saw a human fossil, much less conceived of men different from living men. Between his time and ours the use of the Linnaean system of classification as applied to man and his relatives past and present became almost a game. On grasping the concept of evolution, scholars saw that modern man must have had ancestors. They were prepared to anticipate the actual discovery of these ancestral forms, and perhaps the greatest anticipator was the German biologist Ernst Haeckel. Working on the basis of fragmentary information in 1889, when the only well-known fossil human remains were the comparatively recent bones discovered 25 years earlier in the Neander Valley of Germany, Haeckel drew up a theoretical ancestral line for man. The line began among some postu-

lated extinct apes of the Miocene epoch and reached *Homo sapiens* by way of an imagined group of "ape-men" (Pithecanthropi) and a group of more advanced but still speechless early men (Alali) whom he visualized as the worldwide stock from which modern men had evolved [*see illustration on page 86*]. A creature combining these various presapient attributes took form in the pooled imagination of Haeckel and his compatriots August Schleicher and Gabriel Max. Max produced a family portrait, and the still-to-be-discovered ancestor was given the respectable Linnaean name *Pithecanthropus alalus*.

Were he living today Haeckel would never do such a thing. It is now the requirement of the International Code of Zoological Nomenclature that the naming of any new genus or species be supported by publication of the specimen's particulars together with a description showing it to be recognizably different from any genus or species previously known. Haeckel was rescued from retroactive embarrassment, however, by Dubois, who gave Haeckel's genus name to Java man. The skull was too large to be an ape's and apparently too small to be a man's; the name *Pithecanthropus* seemed perfectly appropriate. On the other hand, the thighbone from the same formation was essentially modern; its possessor had evidently walked upright. Dubois therefore gave his discovery the species name *erectus*. Since Dubois's time the legitimacy of his finds has been confirmed by the discovery in Java (by G. H. R. von Koenigswald between 1936 and 1939 and by Indonesian workers within the past three years) of equally old and older fossils of the same population.

In the 50 years between Dubois's discovery and the beginning of World

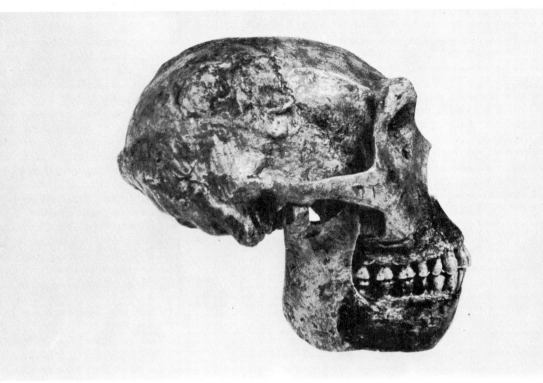

JAVA MAN, whose 700,000-year-old remains were unearthed in 1891 by Eugène Dubois, is representative of the earliest *Homo erectus* population so far discovered. This reconstruction was made recently by G. H. R. von Koenigswald and combines the features of the more primitive members of this species of man that he found in the lowest (Djetis) fossil strata at Sangiran in central Java during the 1930's. The characteristics that are typical of *Homo erectus* include the smallness and flatness of the cranium, the heavy brow-ridge and both the sharp bend and the ridge for muscle attachment at the rear of the skull. The robustness of the jaws adds to the species' primitive appearance. In most respects except size, however, the teeth of *Homo erectus* resemble those of modern man.

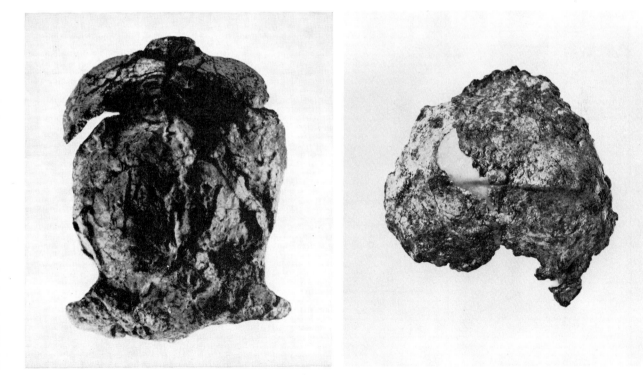

LANTIAN MAN is the most recently found *Homo erectus* fossil. The discovery consists of a jawbone and this skullcap (*top view, browridge at bottom*) from which the occipital bone (*top*) is partially detached. Woo Ju-kang of the Chinese Academy of Sciences in Peking provided the photograph; this fossil man from Shensi may be as old as the earliest specimens of *Homo erectus* from Java.

OCCIPITAL BONE found at Vértesszöllös in Hungary in 1965 is 500,000 or more years old. The only older human fossil in Europe is the Heidelberg jaw. The bone forms the rear of a skull; the ridge for muscle attachment (*horizontal line*) is readily apparent. In spite of this primitive feature and its great age, the skull fragment from Vértesszöllös has been assigned to the species *Homo sapiens*.

War II various other important new kinds of human fossil came into view. For our purposes the principal ones (with some of the Linnaean names thrust on them) were (1) the lower jaw found at Mauer in Germany in 1907 (*Homo heidelbergensis* or *Palaeanthropus*), (2) the nearly complete skull found at Broken Hill in Rhodesia in 1921 (*Homo rhodesiensis* or *Cyphanthropus*), (3) various remains uncovered near Peking in China, beginning with one tooth in 1923 and finally comprising a collection representing more than 40 men, women and children by the end of 1937 (*Sinanthropus pekinensis*), and (4) several skulls found in 1931 and 1932 near

Ngandong on the Solo River not far from where Dubois had worked (*Homo soloensis* or *Javanthropus*). This is a fair number of fossils, but they were threatened with being outnumbered by the names assigned to them. The British student of early man Bernard G. Campbell has recorded the following variants in the case of the Mauer jawbone alone: *Palaeanthropus heidelbergensis, Pseudhomo heidelbergensis, Protanthropus heidelbergensis, Praehomo heidelbergensis, Praehomo europaeus, Anthropus heidelbergensis, Maueranthropus heidelbergensis, Europanthropus heidelbergensis* and *Euranthropus*.

Often the men responsible for these

redundant christenings were guilty merely of innocent grandiloquence. They were not formally declaring their conviction that each fossil hominid belonged to a separate genus, distinct from *Homo*, which would imply an enormous diversity in the human stock. Nonetheless, the multiplicity of names has interfered with an understanding of the evolutionary significance of the fossils that bore them. Moreover, the human family trees drawn during this period showed a fundamental resemblance to Haeckel's original venture; the rather isolated specimens of early man were stuck on here and there like Christmas-tree ornaments. Although the arrangements evinced a vague consciousness of evolution, no scheme was presented that intelligibly interpreted the fossil record.

At last two questions came to the fore. First, to what degree did the fossils really differ? Second, what was the difference among them over a period of time? The fossil men of the most recent period—those who had lived between roughly 100,000 and 30,000 years ago—were Neanderthal man, Rhodesian man and Solo man. They have been known traditionally as *Homo neanderthalensis, Homo rhodesiensis* and *Homo soloensis,* names that declare each of the three to be a separate species, distinct from one another and from *Homo sapiens.* This in turn suggests that if Neanderthal and Rhodesian populations had come in contact, they would probably not have interbred. Such a conclusion is difficult to establish on the basis of fossils, particularly when they are few and tell very little about the geographical range of the species. Today's general view is a contrary one. These comparatively recent fossil men, it is now believed, did not constitute separate species. They were at most incipient species, that is, subspecies or variant populations that had developed in widely separated parts of the world but were still probably able to breed with one another or with *Homo sapiens.*

It was also soon recognized that the older Java and Peking fossils were not very different from one another. The suggestion followed that both populations be placed in a single genus (*Pithecanthropus*) and that the junior name (*Sinanthropus*) be dropped. Even this, however, was one genus too many for Ernst Mayr of Harvard University. Mayr, whose specialty is the evolutionary basis of biological classification, declared that ordinary zoological standards would not permit Java and Peking man to occupy

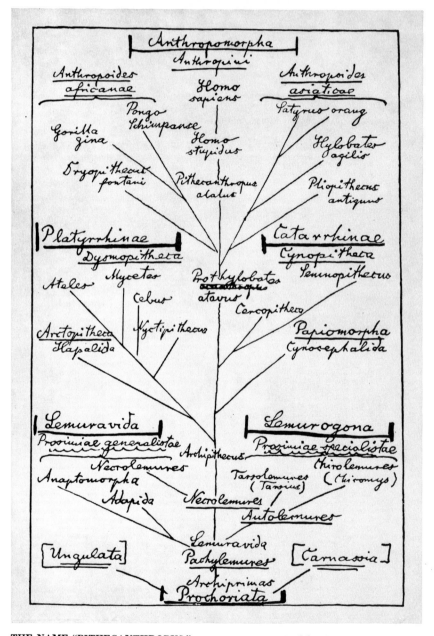

THE NAME "PITHECANTHROPUS," or ape-man, was coined by the German biologist Ernst Haeckel in 1889 for a postulated precursor of *Homo sapiens.* Haeckel placed the ape-man genus two steps below modern man on his "tree" of primate evolution, adding the species name *alalus,* or "speechless," because he deemed speech an exclusively human trait.

GRADE	EUROPE	NORTH AFRICA	EAST AFRICA	SOUTH AFRICA	EAST ASIA	SOUTHEAST ASIA
(5)	*HOMO SAPIENS* (VERTESSZÖLLOS)					
(4)						*(HOMO ERECTUS SOLOENSIS)*
3	*HOMO ERECTUS HEIDELBERGENSIS*	*HOMO ERECTUS MAURITANICUS*	*HOMO ERECTUS LEAKEYI*		*HOMO ERECTUS PEKINENSIS*	
2						*HOMO ERECTUS ERECTUS*
1			*HOMO ERECTUS HABILIS*	*HOMO ERECTUS CAPENSIS*	*(HOMO ERECTUS LANTIANENSIS)*	*HOMO ERECTUS MODJOKERTENSIS*

EIGHT SUBSPECIES of *Homo erectus* that are generally accepted today have been given appropriate names and ranked in order of evolutionary progress by the British scholar Bernard G. Campbell. The author has added Lantian man to Campbell's lowest *Homo erectus* grade and provided a fourth grade to accommodate Solo man, a late but primitive survival. The author has also added a fifth grade for the *Homo sapiens* fossil from Vértesszöllös (*color*). Colored area suggests that Heidelberg man is its possible forebear.

a genus separate from modern man. In his opinion the amount of evolutionary progress that separates *Pithecanthropus* from ourselves is a step that allows the recognition only of a different species. After all, Java and Peking man apparently had bodies just like our own; that is to say, they were attacking the problem of survival with exactly the same adaptations, although with smaller brains. On this view Java man is placed in the genus *Homo* but according to the rules retains his original species name and so becomes *Homo erectus*. Under the circumstances Peking man can be distinguished from him only as a subspecies: *Homo erectus pekinensis*.

The simplification is something more than sweeping out a clutter of old names to please the International Commission on Zoological Nomenclature. The reduction of fossil hominids to not more than two species and the recognition of *Homo erectus* has become increasingly useful as a way of looking at a stage of human evolution. This has been increasingly evident in recent years, as human fossils have continued to come to light and as new and improved methods of dating them have been developed. It is now possible to place both the old discoveries and the new ones much more precisely in time, and that is basic to establishing the entire pattern of human evolution in the past few million years.

To consider dating first, the period during which *Homo erectus* flourished occupies the early middle part of the Pleistocene epoch. The evidence that now enables us to subdivide the Pleistocene with some degree of confidence is of several kinds. For example, the fossil animals found in association with fossil men often indicate whether the climate of the time was cold or warm. The comparison of animal communities is also helpful in correlating intervals of time on one continent with intervals on another. The ability to determine absolute dates, which makes possible the correlation of the relative dates derived from sequences of strata in widely separated localities, is another significant development. Foremost among the methods of absolute dating at the moment is one based on the rate of decay of radioactive potassium into argon. A second method showing much promise is the analysis of deep-sea sediments; changes in the forms of planktonic life embedded in samples of the bottom reflect worldwide temperature changes. When the absolute ages of key points in sediment sequences are determined by physical or chemical methods, it ought to be possible to assign dates to all the major events of the Pleistocene. Such methods have already suggested that the epoch began more than three million years ago and that its first major cold phase (corresponding to the Günz

glaciation of the Alps) may date back to as much as 1.5 million years ago. The period of time occupied by *Homo erectus* now appears to extend from about a million years ago to 500,000 years ago in terms of absolute dates, or from some time during the first interglacial period in the Northern Hemisphere to about the end of the second major cold phase (corresponding to the Mindel glaciation of the Alps).

On the basis of the fossils found before World War II, with the exception of the isolated and somewhat peculiar Heidelberg jaw, *Homo erectus* would have appeared to be a human population of the Far East. The Java skulls, particularly those that come from the lowest fossil strata (known as the Djetis beds), are unsurpassed within the entire group in primitiveness. Even the skulls from the strata above them (the Trinil beds), in which Dubois made his original discovery, have very thick walls and room for only a small brain. Their cranial capacity probably averages less than 900 cubic centimeters, compared with an average of 500 c.c. for gorillas and about 1,400 c.c. for modern man. The later representatives of Java man must be more than 710,000 years old, because potassium-argon analysis has shown that tektites (glassy stones formed by or from meteorites) in higher strata of the same formation are of that age.

The Peking fossils are younger, probably dating to the middle of the second Pleistocene cold phase, and are physically somewhat less crude than the Java ones. The braincase is higher, the face shorter and the cranial capacity approaches 1,100 c.c., but the general construction of skull and jaw is similar. The teeth of both Java man and Peking man are somewhat larger than modern man's and are distinguished by traces of an enamel collar, called a cingulum, around some of the crowns. The latter is an ancient and primitive trait in man and apes.

Discoveries of human fossils after World War II have added significantly to the picture of man's distribution at this period. The pertinent finds are the following:

1949: Swartkrans, South Africa. Jaw and facial fragments, originally given the name *Telanthropus capensis*. These were found among the copious remains at this site of the primitive subhumans known as australopithecines. The fossils were recognized at once by the late Robert Broom and his colleague John T. Robinson as more advanced than the australopithecines both in size and in traits of jaw and teeth. Robinson has now assigned *Telanthropus* to *Homo erectus*, since that is where he evidently belongs.

1955: Ternifine, Algeria. Three jaws and a parietal bone, given the name *Atlanthropus mauritanicus*, were found under a deep covering of sand on the clay floor of an ancient pond by Camille Arambourg. The teeth and jaws show a strong likeness to the Peking remains.

1961: Olduvai Gorge, Tanzania. A skullcap, not formally named but identified as the Bed II Hominid, was discovered by L. S. B. Leakey. Found in a context with a provisional potassium-argon date of 500,000 years ago, the skull's estimated cranial capacity is 1,000 c.c. Although differing somewhat in detail, it has the general characteristics of the two Far Eastern subspecies of *Homo erectus*. At lower levels in this same important site were found the remains of a number of individuals with small skulls, now collectively referred to as "Homo habilis."

1963–1964: Lantian district, Shensi, China. A lower jaw and a skullcap were found by Chinese workers at two separate localities in the district and given the name *Sinanthropus lantianensis*. Animal fossils indicate that the Lantian sites are older than the one that yielded Peking man and roughly as old as the lowest formation in Java. The form of the skull and jaw accords well with this

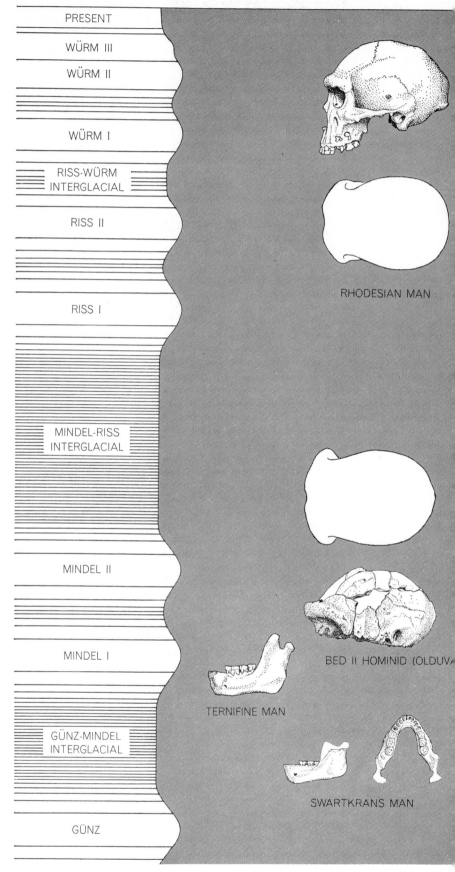

PRESENT
WÜRM III
WÜRM II
WÜRM I
RISS-WÜRM INTERGLACIAL
RISS II
RISS I
MINDEL-RISS INTERGLACIAL
MINDEL II
MINDEL I
GÜNZ-MINDEL INTERGLACIAL
GÜNZ

RHODESIAN MAN

BED II HOMINID (OLDUVA

TERNIFINE MAN

SWARTKRANS MAN

FOSSIL EVIDENCE for the existence of a single species of early man instead of several species and genera of forerunners of *Homo sapiens* is presented in this array of individual remains whose age places them in the interval of approximately 500,000 years that separates the first Pleistocene interglacial period from the end of the second glacial period (*see scale at left*). The earliest *Homo erectus* fossils known, from Java and China, belong to the first interglacial period; the earliest *Homo erectus* remains from South Africa may be equally

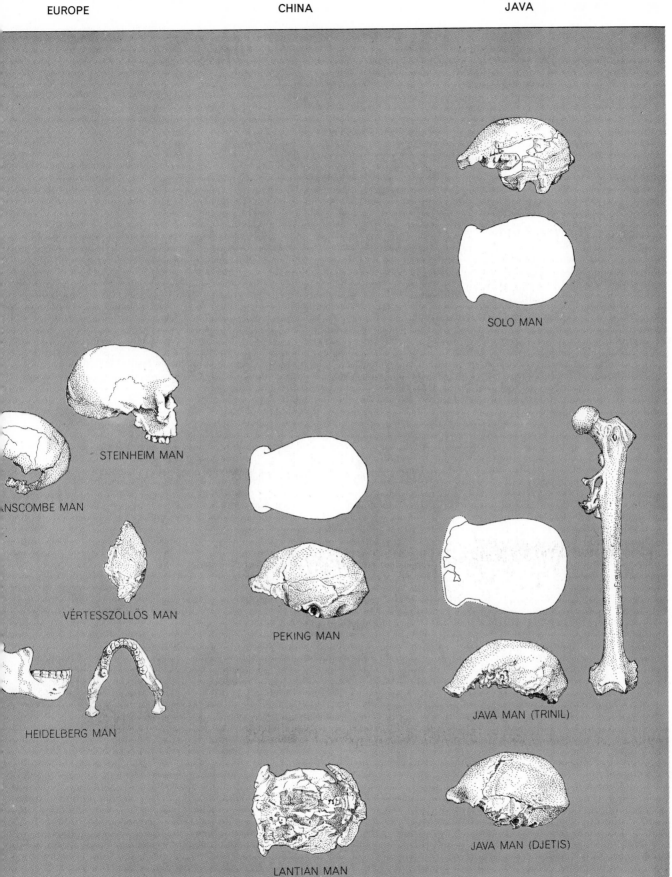

SOLO MAN

STEINHEIM MAN

NSCOMBE MAN

VÉRTESSZOLLÖS MAN

PEKING MAN

JAVA MAN (TRINIL)

HEIDELBERG MAN

LANTIAN MAN

JAVA MAN (DJETIS)

old. Half a million years later *Homo erectus* continued to be represented in China by the remains of Peking man and in Africa by the skull from Olduvai Gorge. In the intervening period this small-brained precursor of modern man was not the only human species inhabiting the earth, nor did *Homo erectus* become extinct when the 500,000-year period ended. One kind of man who had apparent-ly reached the grade of *Homo sapiens* in Europe by the middle or later part of the second Pleistocene glacial period was unearthed recently at Vértesszöllös in Hungary. In the following inter-glacial period *Homo sapiens* is represented by the Steinheim and Swanscombe females. Solo man's remains indicate that *Homo erectus* survived for several hundred thousand years after that.

dating; both are distinctly more primitive than the Peking fossils. Both differ somewhat in detail from the Java subspecies of *Homo erectus*, but the estimated capacity of this otherwise large skull (780 c.c.) is small and close to that of the earliest fossil cranium unearthed in Java.

1965: Vértesszöllös, Hungary. An isolated occipital bone (in the back of the skull) was found by L. Vértes. This skull fragment is the first human fossil from the early middle Pleistocene to be unearthed in Europe since the Heidelberg jaw. It evidently dates to the middle or later part of the Mindel glaciation and thus falls clearly within the *Homo erectus* time zone as defined here. The bone is moderately thick and shows a well-defined ridge for the attachment of neck muscles such as is seen in all the *erectus* skulls. It is unlike *erectus* occipital bones, however, in that it is both large and definitely less angled; these features indicate a more advanced skull.

In addition to these five discoveries, something else of considerable importance happened during this period. The Piltdown fraud, perpetrated sometime before 1912, was finally exposed in 1953. The detective work of J. S. Weiner, Sir Wilfrid Le Gros Clark and Kenneth Oakley removed from the fossil record a supposed hominid with a fully apelike jaw and manlike skull that could scarcely be fitted into any sensible evolutionary scheme.

From this accumulation of finds, many of them made so recently, there emerges a picture of men with skeletons like ours but with brains much smaller, skulls much thicker and flatter and furnished with protruding brows in front and a marked angle in the rear, and with teeth somewhat larger and exhibiting a few slightly more primitive traits. This picture suggests an evolutionary level, or grade, occupying half a million years of human history and now seen

to prevail all over the inhabited Old World. This is the meaning of *Homo erectus*. It gives us a new foundation for ideas as to the pace and the pattern of human evolution over a critical span of time.

Quite possibly this summary is too tidy; before the 100th anniversary of the resurrection of *Homo erectus* is celebrated complications may appear that we cannot perceive at present. Even today there are a number of fringe problems we cannot neglect. Here are some of them.

What was the amount of evolution taking place within the *erectus* grade? There is probably a good deal of accident of discovery involved in defining *Homo erectus*. Chance, in other words, may have isolated a segment of a continuum, since finds from the time immediately following this 500,000-year period are almost lacking. It seems likely, in fact practically certain, that real evolutionary progress was taking place,

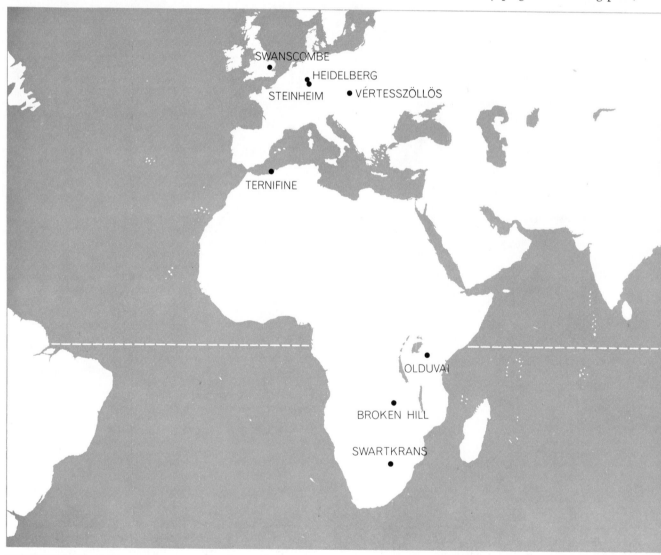

DISTRIBUTION of *Homo erectus* seemed to be confined mainly to the Far East and Southeast Asia on the basis of fossils unearthed

before World War II; the sole exception was the Heidelberg jaw. Postwar findings in South, East and North Africa, as well as dis-

but the tools made by man during this period reveal little of it. As for the fossils themselves, the oldest skulls—from Java and Lantian—are the crudest and have the smallest brains. In Java, one region with some discernible stratigraphy, the later skulls show signs of evolutionary advance compared with the earlier ones. The Peking skulls, which are almost certainly later still, are even more progressive. Bernard Campbell, who has recently suggested that all the known forms of *Homo erectus* be formally recognized as named subspecies, has arranged the names in the order of their relative progressiveness. I have added some names to Campbell's list; they appear in parentheses in the illustration on page 87. As the illustration indicates, the advances in grade seem indeed to correspond fairly well with the passage of time.

What are the relations of *Homo erectus* to Rhodesian and Solo man? This is a point of particular importance, be-

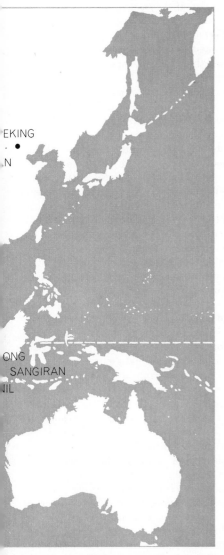

covery of a new *Homo erectus* site in northern China, have extended the species' range.

cause both the African and the Javanese fossils are much younger than the date we have set as the general upward boundary for *Homo erectus*. Rhodesian man may have been alive as recently as 30,000 years ago and may have actually overlapped with modern man. Solo man probably existed during the last Pleistocene cold phase; this is still very recent compared with the time zone of the other *erectus* fossils described here. Carleton S. Coon of the University of Pennsylvania deems both late fossil men to be *Homo erectus* on the basis of tooth size and skull flatness. His placing of Rhodesian man is arguable, but Solo man is so primitive, so like Java man in many aspects of his skull form and so close to Peking man in brain size that his classification as *Homo erectus* seems almost inevitable. The meaning of his survival hundreds of thousands of years after the period I have suggested, and his relation to the modern men who succeeded him in Southeast Asia in recent times, are unanswered questions of considerable importance.

Where did *Homo erectus* come from? The Swartkrans discovery makes it clear that he arose before the last representatives of the australopithecines had died out at that site. The best present evidence of his origin is also from Africa; it consists of the series of fossils unearthed at Olduvai Gorge by Leakey and his wife and called Homo habilis. These remains seem to reflect a transition from an australopithecine level to an *erectus* level about a million years ago. This date seems almost too late, however, when one considers the age of *Homo erectus* finds elsewhere in the world, particularly in Java.

Where did *Homo erectus* go? The paths are simply untraced, both those that presumably lead to the Swanscombe and Steinheim people of Europe during the Pleistocene's second interglacial period and those leading to the much later Rhodesian and Neanderthal men. This is a period lacking useful evidence. Above all, the nature of the line leading to living man—*Homo sapiens* in the Linnaean sense—remains a matter of pure theory.

We may, however, have a clue. Here it is necessary to face a final problem. What was the real variation in physical type during the time period of *Homo erectus?* On the whole, considering the time and space involved, it does not appear to be very large; the similarity of the North African jaws to those of Peking man, for example, is striking in spite of the thousands of

miles that separated the two populations. The Heidelberg jaw, however, has always seemed to be somewhat different from all the others and a little closer to modern man in the nature of its teeth. The only other European fossil approaching the Heidelberg jaw in antiquity is the occipital bone recently found at Vértesszöllös. This piece of skull likewise appears to be progressive in form and may have belonged to the same general kind of man as the Heidelberg jaw, although it is somewhat more recent in date.

Andor Thoma of Hungary's Kossuth University at Debrecen in Hungary, who has kindly given me information concerning the Vértesszöllös fossil, will publish a formal description soon in the French journal *L'Anthropologie*. He estimates that the cranial capacity was about 1,400 c.c., close to the average for modern man and well above that of the known specimens of *Homo erectus*. Although the occipital bone is thick, it is larger and less sharply angled than the matching skull area of Rhodesian man. It is certainly more modern-looking than the Solo skulls. I see no reason at this point to dispute Thoma's estimate of brain volume. He concludes that Vértesszöllös man had in fact reached the *sapiens* grade in skull form and brain size and accordingly has named him a subspecies of *Homo sapiens*.

Thoma's finding therefore places a population of more progressive, *sapiens* humanity contemporary with the populations of *Homo erectus* 500,000 years ago or more. From the succeeding interglacial period in Europe have come the Swanscombe and Steinheim skulls, generally recognized as *sapiens* in grade. They are less heavy than the Hungarian fossil, more curved in occipital profile and smaller in size; they are also apparently both female, which would account for part of these differences.

The trail of evidence is of course faint, but there are no present signs of contradiction; what we may be seeing is a line that follows *Homo sapiens* back from Swanscombe and Steinheim to Vértesszöllös and finally to Heidelberg man at the root. This is something like the Solo case in reverse, a *Homo sapiens* population surprisingly early in time, in contrast to a possible *Homo erectus* population surprisingly late. In fact, we are seeing only the outlines of what we must still discover. It is easy to perceive how badly we need more fossils; for example, we cannot relate Heidelberg man to any later Europeans until we find some skull parts to add to his solitary jaw.

9

Stone Tools and Human Behavior

by Sally R. Binford and Lewis R. Binford
April 1969

Statistical analysis of the implements found at Paleolithic sites can identify the groups of tools that were used for various kinds of jobs. These groupings suggest how early man's life was organized

The main evidence for almost the entire span of human prehistory consists of stone tools. Over the more than three million years of the Pleistocene epoch hunting and gathering peoples left behind them millions of such tools, ranging from crudely fractured pebbles to delicately flaked pieces of flint. Modern students of these objects are attempting to understand their various functions, and much of current prehistoric research is concerned with developing methods for achieving this understanding.

For many years prehistorians devoted almost all their efforts to establishing cultural sequences in order to determine what happened when. Chronologies have been established for many parts of the Old World, both on the basis of stratigraphy and with the aid of more modern techniques such as radioisotope dating and pollen analysis. Although many details of cultural sequences remain to be worked out, the broad outlines are known well enough for prehistoric archaeologists to address themselves to a different range of questions, not so much what happened when as what differences in stone tools made at the same time mean.

Traditionally differences in assemblages of stone tools from the same general period were thought to signify different cultures. Whereas the term "culture" was never very clearly defined, it most often meant distinct groups of people with characteristic ways of doing things, and frequently it was also taken to mean different ethnic affiliations for the men responsible for the tools. Such formulations cannot readily be tested and so are scientifically unsatisfactory. If we were to examine the debris left behind by people living today, we would find that differences in such material could most often be explained by differences in human activities. For example, the kinds of archaeological remains that would be left by a modern kitchen would differ markedly from those left by miners. This variation in archaeological remains is to be understood in terms of function—what activities were carried out at functionally different locations—and not in terms of "kitchen cultures" or "mining cultures."

The example is extreme, but it serves to illustrate a basic difference between the assumptions underlying our research and those on which the more traditional prehistory is based. The obvious fact that human beings can put different locations to different uses leads us to the concept of settlement type and settlement system, the framework that seems most appropriate for interpreting prehistoric stone-tool assemblages. In what follows we are restating, and in some respects slightly modifying, some useful formulations put forward by Philip L. Wagner of Simon Fraser University in British Columbia.

All known groups of hunter-gatherers live in societies composed of local groups that can be internally organized in various ways; invariably the local group is partitioned into subgroups that function to carry out different tasks. Sex and age are the characteristics that most frequently apply in the formation of subgroups: the subgroups are generally composed of individuals of the same age or sex who cooperate in a work force. For example, young male adults often cooperate in hunting, and women work together in collecting plant material and preparing food. At times a larger local group breaks up along different lines to form reproductive-residence units, and these family subgroups tend to be more permanent and self-sustaining than the work groups.

Although we have no idea how prehistoric human groups were socially partitioned, it seems reasonable to assume that these societies were organized flexibly and included both family and work groups. If the assumption is correct, we would expect this organization to be reflected in differences both between stone-tool assemblages at a given site and between assemblages at different sites.

Geographical variations would arise because not all the activities of a given society are conducted in one place. The ways that game, useful plants, appropriate living sites and the raw materials for tool manufacture are distributed in the environment will directly affect where subgroups of a society perform different activities. One site might be a favorable place for young male hunters to kill and partly butcher animals; another might be a more appropriate place for women and children to gather and process plants. Both work locations might be some distance from the group's main living site. One would expect the composition of the tool assemblages at various locations to be determined by the kinds of tasks performed and by the size and composition of the group performing them.

Temporal variations can also be expected between assemblages of stone tools, for several reasons. The availability of plants and animals in the course of the year is a primary factor; it varies as a result of the reproductive cycles of the plants and animals. The society itself varies in an annual cycle; the ways the members of a society are organized and how they cooperate at different times of the year change with their activities at different seasons. Moreover, any society must solve integrative problems as a result of the maturation of the young, the death of some members, relations with other groups and so on. The behavioral modifications prompted by such considerations will be reflected in the society's use of a territory.

In addition to these factors that can affect the archaeological material left behind by a society, there are other determinants that profoundly modify site utilization. It is the kinds of site used for different activities and the way these specialized locations are related that are respectively termed settlement type and settlement system.

In technologically simple societies we can distinguish two broad classes of activities: extraction and maintenance. Extraction involves the direct procurement of foods, fuels and raw materials for tools. Maintenance activities consist in the preparation and distribution of foods and fuels already on hand and in the processing of raw materials into tools. Since the distribution of resources in the environment is not necessarily related to the distribution of sites providing adequate living space and safety, we would not expect extraction and maintenance activities to be conducted in the same places.

Base camps are chosen primarily for living space, protection from the elements and central location with respect to resources. We would expect the archaeological assemblages of base camps to reflect maintenance activities: the preparation and consumption of food and the manufacture of tools for use in other less permanent sites.

Another settlement type would be a work camp, a site occupied while smaller social units were carrying out extractive tasks. Archaeologically these would appear as kill sites, collecting stations and quarries for extracting flint to be used in toolmaking. The archaeological assemblages from these sites should be dominated by the tools used in the specific extractive tasks. If a work camp were occupied for a rather long period and by a fairly large subgroup, we would anticipate that some maintenance activities would also be reflected in the archaeological remains.

It is the way these two general classes of camps are used by any society that defines the settlement system. If a hunting-gathering society were relatively sedentary, we would expect the tools at the base camp to exhibit little seasonal variation because the base camp would have been occupied for most of the year. Under some conditions, however, we would expect to find more than one kind of base camp for a society. If the organization of

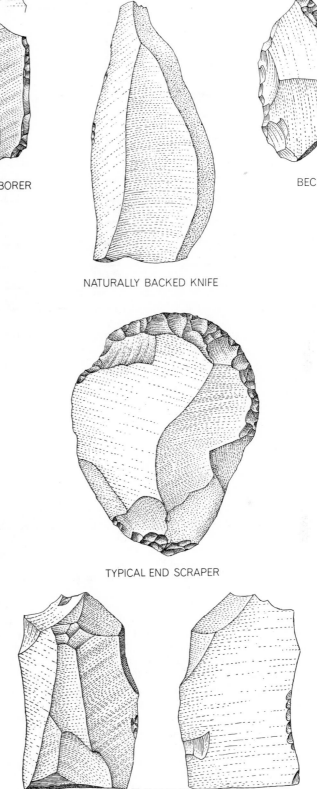

TYPICAL BORER

BEC

NATURALLY BACKED KNIFE

TYPICAL END SCRAPER

ATYPICAL BURIN

FIVE CLASSES OF STONE TOOLS predominate in the largest of the five groups, or factors, revealed by the authors' analyses. Of the 40 classes of tools subjected to multivariate analysis, 16 appear in this cluster, named Factor I. Few of the classes seem suited to hunting or heavy work; they were probably base-camp items for making other tools of wood or bone.

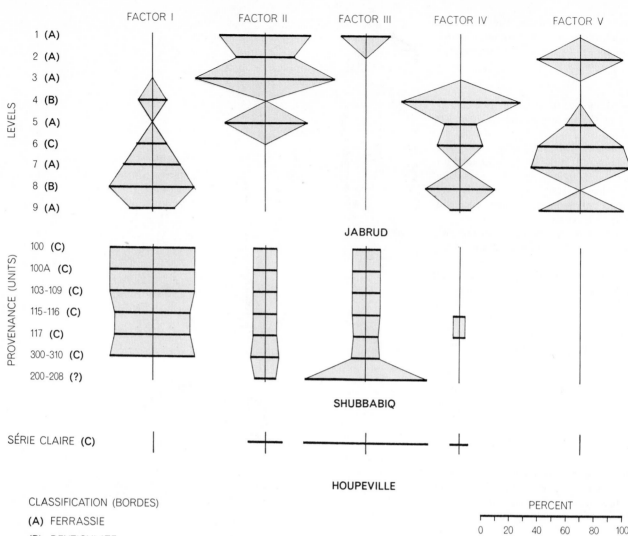

FACTOR I FACTOR II FACTOR III FACTOR IV FACTOR V

LEVELS

1 (A)
2 (A)
3 (A)
4 (B)
5 (A)
6 (C)
7 (A)
8 (B)
9 (A)

JABRUD

PROVENANCE (UNITS)

100 (C)
100A (C)
103-109 (C)
115-116 (C)
117 (C)
300-310 (C)
200-208 (?)

SHUBBABIQ

SÉRIE CLAIRE (C)

HOUPEVILLE

CLASSIFICATION (BORDES)

(A) FERRASSIE
(B) DENTICULATE
(C) TYPICAL

PERCENT

0   20   40   60   80   100

RELATIVE SIGNIFICANCE of the five factors identified in the tool assemblages from three Mousterian sites is indicated by the percentage of total variation attributable to each factor in each sample analyzed. Every sample but one bears the name (*key, bottom left*) by which the French archaeologist François Bordes characterizes the entire assemblage. The homogeneity of the samples from the Mughâret es-Shubbabiq cave site in Israel (*center*) is in sharp contrast to the heterogeneity of the samples from the Jabrud rock-shelter in Syria (*top*). The authors suggest that the cave was a base camp but that the rock-shelter was only a work camp, occupied at different times by work parties with different objectives. The Houpeville assemblage (*bottom*) is not like either of the others.

the society changes during the year, perhaps consisting of larger groups during the summer months and dispersing into smaller family units during the lean winter months, there would be more than one kind of base camp, and each would have its distinct seasonal characteristics.

The work camps would display even greater variation; each camp would be occupied for a shorter time, and the activities conducted there would be more specifically related to the resources being exploited. One must also consider how easy or how difficult it was to transport the exploited resource. If a party of hunters killed some big animals or a large number of smaller animals, the entire group might assemble at the kill site

not only to eat but also to process the large quantities of game for future consumption. In such a work camp we would expect to find many of the kinds of tools used for food processing, even though the tasks undertaken would be less diverse than those at a base camp.

The extent to which maintenance tasks are undertaken at work camps will also be directly related to the distance between work camp and base camp. If the two are close together, we would not find much evidence of maintenance activities at the work camp. As the distance between work camp and base camp increases, however, the work-camp assemblage of tools would reflect an increase in maintenance activities. This leads us to

suggest a third type of settlement: the transient camp. At such a location we would find only the most minimal evidence of maintenance activities, such as might be undertaken by a traveling group in the course of an overnight stay.

We have outlined here the settlement system of technologically simple hunter-gatherers. Although the system is not taken directly from any one specific living group, it does describe the generalized kind of settlement system that ethnographers have documented for people at this level of sociocultural complexity. In order to assess the relevance of such a settlement system for hunter-gatherers in the Paleolithic period it was necessary first to relate stone tools to human activities and then to determine how these tools were distributed at different types of site.

The kind of analysis we carried out might well have been impossible without the basic work on the classification of stone tools done by François Bordes of the University of Bordeaux. The archaeological taxonomy devised by Bordes for the Middle Paleolithic has become a widely accepted standard, so that it is now possible for prehistorians working with Middle Paleolithic materials from different parts of the world to describe the stone tools they excavate in identical and repeatable terms.

In addition to compiling a type list of Mousterian, or Middle Paleolithic, tools, Bordes has offered convincing arguments against the "index fossil" approach to the analysis of stone tools. This approach, borrowed from paleontology, assigns a high diagnostic value to the disappearance of an old tool form or the appearance of a new one; such changes are assumed to indicate key cultural events. Bordes has insisted on describing assemblages of tools in their entirety without any a priori assumption that some tools have greater cultural significance than others. This radical departure in classification, combined with highly refined excavation techniques, provides a sound scientific basis on which much current prehistoric research rests.

According to what has become known as *la méthode Bordes*, stone artifacts are classified according to explicitly stated attributes of morphology and technique of manufacture. The population of stone tools from a site (the assemblage) is then described graphically, and the relative frequencies of different kinds of stone tools from various sites can be compared. Such a statistical technique, which deals with a single class of variables, is quite appropriate for the description of assemblages of stone tools. The explanation of multiple similarities and differences, however, requires different statistical techniques.

The factors determining the range and form of activities conducted by any group at any site may vary in terms of many possible "causes" in various combinations. The more obvious among these might be seasonally regulated phenomena affecting the distribution of game, environmental conditions, the ethnic composition of the group, the size and structure of the group regardless of ethnic affiliation and so on. Other determinants of activities might be the particular situation of the group with respect to

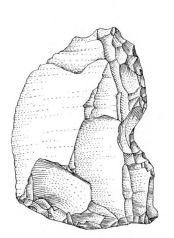

NOTCHED PIECE

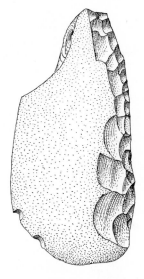

SIDE SCRAPER WITH ABRUPT RETOUCH

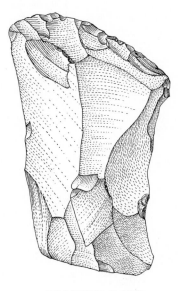

TRUNCATED FLAKE

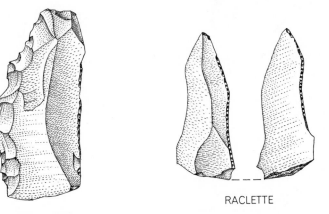

DENTICULATE

RACLETTE

UNIDENTIFIED TOOL KIT, with five predominant classes of artifacts, comprises Factor **IV.** The tasks for which it was intended are not known. Bordes has suggested, however, that denticulates (*bottom left*) may have been utilized for the processing of plant materials.

food, shelter, the supply of tools or the availability of raw materials. In short, the "causes" of assemblage variation are separate activities, each of which is related to the physical and social environment and to the others.

Given this frame of reference, the summary description of frequencies of tool types in an assemblage, which is the end product of Bordes's method, represents a blending of activity units and their determinants. We needed to partition assemblages into groups of tools that reflect activities. To use a chemical analogy, the end product of Bordes's method of describing the entire assemblage is a compound; we hoped to isolate smaller units, analogous to the constituent elements of a compound, that would represent activities. In our view variation in assemblage composition is directly related to the form, nature and spatial arrangement of the activities in which the tools were used.

Since the settlement-system model we had in mind is based on ethnographic examples, we wanted to ensure that the archaeological materials we analyzed were made by men whose psychological capacities were not radically different from our own. The Mousterian, a culture complex named after the site of Le Moustier in the Dordogne, dates from about 100,000 to 35,000 B.C. Mousterian tools are known from western Europe, the Near East, North Africa and even central China. Where human remains have been discovered in association with Mousterian tools they are the remains of Neanderthal man. Once considered to be a species separate from ourselves, Neanderthal man is generally accepted today as a historical subspecies of fully modern man. A great deal of archaeological evidence collected in recent years strongly suggests that the behavioral capacities of Neanderthal man were not markedly different from our own.

The Mousterian assemblages we chose for our analysis came from two sites in the Near East and one in northern France. One of us had excavated a cave site in Israel (Mugharet es-Shubbabiq, near Lake Tiberias) and had analyzed the stone tools found there in Bordes's laboratory and under his supervision. We also used assemblages that had been excavated in the 1930's by Alfred Rust at Jabrud, a rock-shelter near Damascus in Syria; this material had been studied and reclassified by Bordes. The French material came from the open-air site of Houpeville and had been excavated and analyzed by Bordes. We chose these samples because they represented three kinds of site, but more important because they had all been classified by Bordes. This meant that extraneous variation due to the vagaries of classification was eliminated.

To describe the two Near Eastern sites briefly, Shubbabiq is a large cave located in a narrow, deep valley that is dry for most of the year. The cave mouth faces east and its floor covers nearly 350 square meters, with slightly less than 300 meters well exposed to natural light. Un-

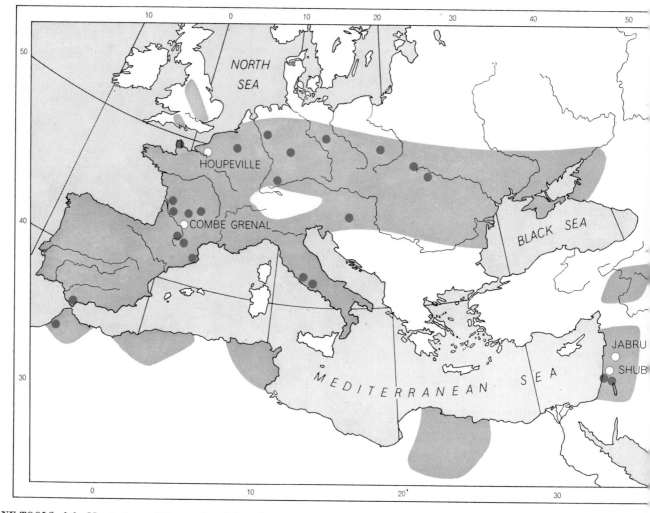

STONE TOOLS of the Mousterian tradition are found throughout Europe and also in the Near East. They were made from 100,000 to 35,000 years ago and are associated in many instances with the remains of Neanderthal man (*colored dots*). The authors' statistical

fortunately the Mousterian deposits in the main part of the cave had been destroyed by more recent inhabitants. Five of the samples from Shubbabiq used in our study came from deposits in the rear of the cave. The sixth, Unit 200–208, was a small deposit outside the cave entrance. The Syrian site, Jabrud Shelter I, is long and narrow. Like Shubbabiq, it faces east, and because it is more open it is much more exposed to the elements. Located on the edge of the Anti-Lebanon range, it looks down on the floor of a valley. At the time of the occupations that interested us the shelter had about 178 square meters of floor space. Rust excavated a trench some 23 meters long and three meters wide along the shelter's back wall. The shelter yielded many layers of Mousterian tools, but only the upper nine strata contained assemblages that could be compared with those from Shubbabiq.

Our study sought to answer three questions. First, does the composition of the total assemblage from any occupa-

analyses utilized tools from two sites in the Near East and two in Europe (*open circles*).

tion level correspond to any single human activity, or does information summarized in a single class of variables obscure the fact that each assemblage represents an assortment of activities? Second, is there any regularity in the composition of assemblages at a single location that can be interpreted in terms of regular patterns of human behavior? Third, is there some kind of directional change over a period of time in assemblages from a single location that suggests evolutionary changes in human behavior?

The statistical analysis of a single class of variables is termed univariate analysis. Multivariate statistical analysis allows one to calculate the measure of dependence among many classes of variables in several ways. Because we needed to determine the measure of dependence relating every one of some 40 classes of tools found in varying percentages in 17 different assemblages from three sites to every one of the remaining 39 classes, we faced a staggering burden of calculations. Such a job could not have been undertaken before the advent of high-speed computers. Factor analysis, which has been applied in areas as unrelated as geology and sociology, seemed the most appropriate method. Our analysis was run at the University of Chicago's Institute for Computer Research, with the aid of a modification of a University of California program for factor analysis (Mesa 83) and an IBM 7090 computer.

The factor analysis showed that in our samples the different classes of Mousterian artifacts formed five distinct clusters. The most inclusive of the five—Factor I—consists of 16 types of tools. Within this grouping the tools showing the highest measure of mutual dependence are, to use Bordes's taxonomic terminology, the "typical borer," the "typical end scraper," the "bec" (a small, beaked flake), the "atypical burin" (an incising tool) and the "naturally backed knife" [*see illustration on page 93*]. None of these tools is well suited to hunting or to heavy-duty butchering, but most are well designed for cutting and incising wood or bone. (The end scraper seems best adapted to working hides.) On these grounds we interpret Factor I as representing activities conducted at a base camp.

The next grouping produced by the factor analysis we interpret as a kit of related tools for hunting and butchering. The tools in this group—Factor II—are of 12 types. Three varieties of spear point are dominant; in Bordes's terminology they are the "plain Levallois point," the

"retouched Levallois point" and the "Mousterian point." Side scrapers of four classes are the other tools that show the highest measure of mutual dependence: the "simple straight," the "simple convex," the "convergent" and the "double" side scraper [*see illustration on page 18*].

In Factor III the main diagnostic tools are cutting implements. They include the "typical backed knife," the "naturally backed knife" (also found in Factor I), "typical" and "atypical" Levallois flakes, "unretouched blades" and "end-notched pieces" [*see illustration on page 97*]. With the exception of the end-notched pieces all these tools appear to be implements for fine cutting. Their stratigraphic association with evidence of fire suggests that Factor III is a tool kit for the preparation of food.

Factor IV is distinctive. Its characteristic tools are "denticulates" (flakes with at least one toothed edge), "notched pieces," "side scrapers with abrupt retouch," "raclettes" (small flakes with at least one delicately retouched edge) and "truncated flakes." We find it difficult even to guess at the function of this factor. Bordes has suggested that some of these tools were employed in the processing of plant materials.

The tools with the highest measure of mutual dependence in Factor V are "elongated Mousterian points," "disks," "scrapers made on the ventral surfaces of flakes," "typical burins" (as opposed to the atypical burin in Factor I) and "unretouched blades" (which are also found in Factor III). The fact that there is only one kind of point and one kind of scraper among the diagnostic implements suggests that Factor V is a hunting and butchering tool kit that is more specialized than the one represented by Factor II.

What answers does the existence of five groups of statistically interdependent artifacts among Mousterian assemblages give to the three questions we raised? In response to the first question we can show that neither at Shubbabiq nor at Jabrud does the total assemblage correspond to any single human activity. The degree to which individual factors account for the variation between assemblages can be expressed in percentages [*see illustration on page 14*]. To consider the Shubbabiq findings first, the percentages make it plain that, with the exception of a group of tools in Unit 200–208, the assemblages as a whole are internally quite consistent. The major part of the variation is accounted for by

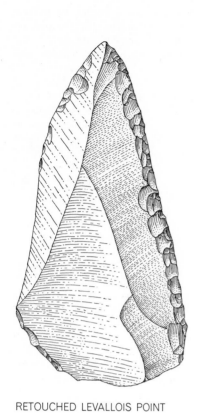

RETOUCHED LEVALLOIS POINT

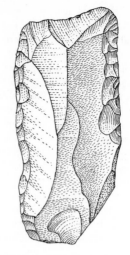

LEVALLOIS POINT

DOUBLE SIDE SCRAPER

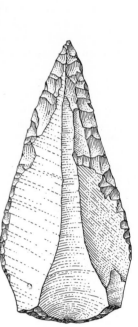

MOUSTERIAN POINT

CONVERGENT SIDE SCRAPER

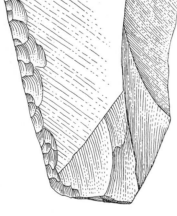

STRAIGHT SIDE SCRAPER

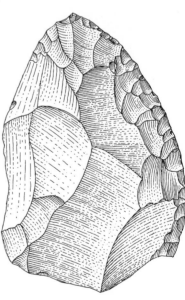

CONVEX SIDE SCRAPER

POINTS AND SCRAPERS outnumber other kinds of tools among the 12 classes comprising the second-largest factor. The seven predominant classes in the assemblage are illustrated. Factor II is evidently an assemblage suited to hunting and butchering animals. This illustration and the others of Mousterian tools are based on original drawings by Pierre Laurent of the University of Bordeaux.

Factor I, the base-camp grouping; the remainder is shared between Factor II, the all-purpose hunting and butchering tool kit, and Factor III, the food-preparation cluster. The distinctive denticulate factor—Factor IV—appears in only two samples and represents less than 10 percent of the variability in each.

As we have mentioned, Unit 200–208 consists of tools from the deposit outside the cave mouth. Three small accumulations of ash—evidence of fires—were also found in the deposits. It seems more than coincidence that the tool grouping dominating this unit is the one associated with food preparation. In any event, the consistent homogeneity of the other excavation units at Shubbabiq suggests that the cave served the same purpose throughout its occupancy, a finding that also answers our second question. Although the occupation of the cave may have spanned a considerable period of time, the regularity of the factors suggests a similar regularity in the behavior of the occupants.

The percentages of variability accounted for at the Jabrud rock-shelter suggest in turn that Jabrud served repeatedly as a work camp where hunting was the principal activity. Evidently the valley the shelter overlooks was rich in game. The animal bones collected during the original excavation of the site have been lost, but recent work at the same site by Ralph S. Solecki of Columbia University indicates that the valley once abounded in horses—herd animals that were frequently killed by Paleolithic hunters.

The Jabrud findings also provide an answer to our third question. The decreasing importance of Factor V—the specialized hunting tool kit—and its replacement by the more generalized hunting equipment of Factor II suggests directional changes in the behavior of Jabrud's inhabitants. The same is true of the steady decline and eventual disappearance of the base-camp maintenance tools represented by Factor I.

Some of the data from Jabrud even provide a hint of a division of labor by sex in the Middle Paleolithic. The tools characteristic of Factor IV are quite consistently made of kinds of flint that are available in the immediate vicinity of the site, whereas the hunting tools tend to be made of flint from sources farther away. If, in accordance with Bordes's suggestion, the denticulates were used primarily to process plant materials, the expedient fashioning of denticulates out of raw materials on the spot coincides nicely with the fact that

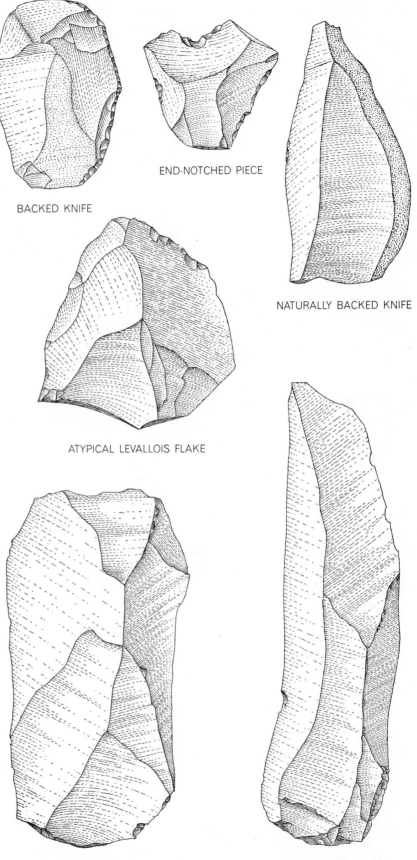

BACKED KNIFE

END-NOTCHED PIECE

NATURALLY BACKED KNIFE

ATYPICAL LEVALLOIS FLAKE

TYPICAL LEVALLOIS FLAKE

UNRETOUCHED BLADE

TOOLS FOR FINE CUTTING are the predominant implements of Factor III. An exception (*top middle*) belongs to the class of end-notched pieces. Their association with hearths suggests that the knives, blades and flakes of Factor III were used for food preparation.

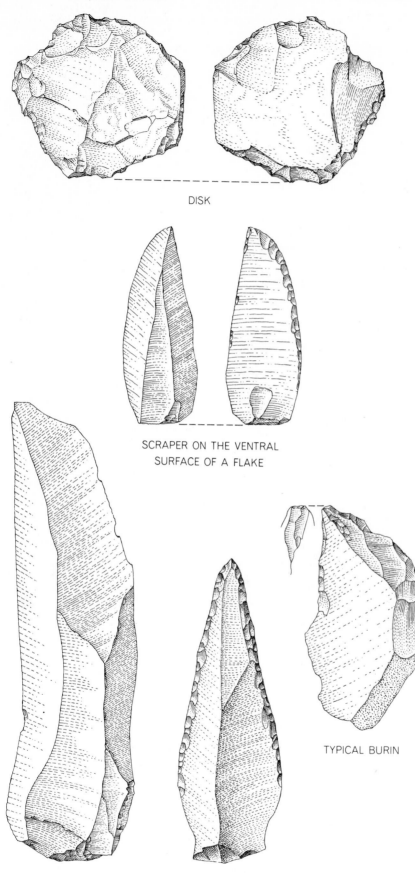

DISK

SCRAPER ON THE VENTRAL
SURFACE OF A FLAKE

TYPICAL BURIN

UNRETOUCHED BLADE   ELONGATED MOUSTERIAN POINT

MORE HUNTING TOOLS are found in Factor V; the five predominant classes are illustrated. The presence of only one class of points and one of scrapers suggests, however, that Factor V reflects specialized hunting rather than the general hunting implied by Factor II.

among living hunter-gatherers women are responsible for the collecting and processing of plant materials.

Recent advances in understanding of the minimum number of persons needed to maintain a self-sustaining human social unit provides additional evidence in favor of our view that Jabrud served as a work camp and that Shubbabiq was a base camp. William W. Howells of Harvard University has suggested that a self-sustaining group must number between 20 and 24 individuals. (He does not imply that the group would necessarily remain together during the entire year.) Taking Howells' estimate as a starting point, we can propose that any base camp where a group could live at full strength must include enough space for the daily activities of 20 to 24 people over a period of several months. Raoul Naroll of the State University of New York at Buffalo suggests that the minimum amount of sheltered space required by an individual is some 10 square meters. On this basis the 178 square meters of floor space in the Jabrud shelter could not have accommodated more than 18 individuals. Shubbabiq cave has enough sheltered floor space for 25 to 30 individuals. Taken together with the differences in the composition of the tool assemblages at the two sites, this leads us to conclude that the sites are basically different types of settlement within a differentiated settlement system.

The tools from Houpeville, called the *Série Claire*, has a totally different geographical context. Since Houpeville is an open-air site and the only one in the study, we feel that an attempt to draw conclusions about its function would be almost meaningless. The *Série Claire* sample nonetheless offers a further demonstration of the power of multivariate analysis. When calculated by univariate statistics, the total configuration of the Houpeville assemblage strongly resembles that of Shubbabiq, that is, the summarized statistics of frequencies of tool types are very similar. When subjected to factor analysis, however, the assemblages from the two sites look quite different. Factors I and V are missing altogether at Houpeville; Factor III, the cluster of food-preparation implements that constitutes a minor percentage of the variability at Shubbabiq, is the major component of the *Série Claire*.

Although we found the results of this factor analysis provocative, it was quite clear that many of our specific interpretations of the factors could not be tested on the basis of such limited data. We felt it essential to add other classes of information to the analysis: animal bones, pollen counts (as checks on climatic inferences) and the distribution of other traces of man (such as hearths) within each occupation level. Such data were available from the site of Combe Grenal, a deeply stratified rock-shelter in the Dordogne region of France excavated by Bordes. They are undoubtedly the finest and most complete Mousterian data in the world. Soil analysis has been done of all the deposits; animal bones are well preserved; pollen profiles have been made for all the 55 Mousterian occupation levels. The sophisticated excavation techniques used at Combe Grenal make it possible to reconstruct the relation of each tool at the site to other tools, to hearths and to clusters of animal bones. We were thus delighted when Bordes graciously volunteered to allow us to analyze his findings.

Our analysis of the Combe Grenal data has occupied the past eight months. (It has been made possible by a grant from the National Science Foundation.) While the work is far from complete, results of a preliminary factor analysis can be summarized here. First, the larger and more complete sample has shown a far wider range of variability than the smaller samples from the Near East have. The tool assemblages in all the Mousterian levels thus far analyzed—41 in number—consist of two or more factors. The factor analysis produced a total of 14 distinct tool groupings, in contrast to the five factors in the Near Eastern sites. In comparing the Combe Grenal analysis with that of the material from the Near East we note some gratifying

consistencies. Such a replication of results with independent data from another region suggests that we have managed to isolate tool groupings that have genuine behavioral significance.

In attempting to relate clusters of tool types to environmental variables such as climate (measured by pollen and sediment studies) and game (as shown by animal bones) we have found no simple, direct form of correlation. There is, however, a nonrandom distribution of the frequency with which a given factor appears in levels that are representative of different environments. It appears that major shifts in climate, sufficient to cause shifts in the distribution of plants and animals, did precipitate a series of adaptive readjustments among the inhabitants of Combe Grenal.

Our present work on the material from Combe Grenal has led us to propose a series of refinements in interpretation. It is clear, for example, that the portability of game played a significant role in determining whether an animal was butchered where it was killed or after it was carried back to the site. We are now reclassifying the bones from that site by categories based on size, as well as by anatomical parts represented, and this should provide information that is not currently discernible. Whether an animal is an upland or a valley form and whether it occurs as one of a herd or as an individual is also evidently significant. We suggest that the behavior of the animals hunted had a profound effect on the degree of preparation for the hunt and on the size and composition of the hunting groups.

It should be stressed that the findings presented here are our own and not Bordes's. As a matter of fact, discussions

of our interpretations with Bordes are usually lively and sometimes heated, although they are always useful. We all agree that Combe Grenal contains so much information in terms of so many different and independent classes of data that many kinds of hypothesis can be tested. Indeed, a procedure that requires the testing and retesting of every interpretation against independent classes of data could be the most significant outcome of our work.

If one goal of prehistory is the accurate description of past patterns of life, certainly it is the job of the archaeologist to explain the variability he observes. Explanation, however, involves the formulation and testing of hypotheses rather than the mere assertion of the meaning of differences and similarities. Many traditionalists speak of "reading the archaeological record," asserting that facts speak for themselves and expressing a deep mistrust of theory. Facts never speak for themselves, and archaeological facts are no more articulate than those of physics or chemistry. It is time for prehistory to deal with the data according to sound scientific procedure. Migrations and invasions, man's innate desire to improve himself, the relation of leisure time to fine arts and philosophy—these and other unilluminating clichés continue to appear in the literature of prehistory with appalling frequency. Prehistory will surely prove a more fruitful field of study when man is considered as one component of an ecosystem—a culture-bearing component, to be sure, but one whose behavior is rationally determined.

10

The Functions of Paleolithic Flint Tools

by Lawrence H. Keeley
November 1977

The microscopic examination of the working edges of certain stone implements used by ancient hunters makes it possible to distinguish among such uses as scraping hide, cutting meat and sawing wood

Almost the only evidence of man's presence on the earth for a period of more than half a million years is vast numbers of stone tools. Some are made of basalt, some of quartzite or quartz and some of the volcanic glass obsidian. In many places the majority are made of flint. As soon as these objects were recognized as man's handiwork they were assigned names based on guesses about their probable function. The French began the process with *coup-de-poing,* which in English became "hand axe." A multitude of other functional names followed: "end scraper," "side scraper," "blade," "point," "burin" and the like. Although generations of prehistorians have used such names, there has been scarcely any tangible evidence on what purposes the stone tools actually served.

Over the past 15 years students of early man have grown sufficiently dissatisfied with this state of affairs to do something about it. The result has been the development of a methodology known as microwear analysis, which reveals the functions of many early flint implements. The evidence is almost indelibly recorded in the form of microscopic traces of wear on the working edges of the flint.

One reason for the current lively interest in the function of stone tools is that progress in the methods of absolute dating, such as carbon-14 analysis, has freed many prehistorians from two former preoccupations. The first was, in the absence of absolute dating, the construction of relative chronologies. The second was closely related to the first: it was the search for "cultural" similarities between assemblages of stone tools from different areas. Such similarities aid in the construction of interlocking regional chronologies. Early in the 1960's a new school of prehistorians began to offer fresh hypotheses to explain the variations between and within regional assemblages of tools.

In this view the variations were attributable less to chronological and cultural differences and more to differences in function. For example, the new school sought to explain the differences between the kinds of tools present in two roughly contemporaneous assemblages in terms of the different kinds of activity the tools' users could have pursued in the two places. Proponents of this school argued that in attributing such differences to "cultural" distinctions between two unrelated groups the older school was misreading the evidence.

A vital prerequisite to the testing of the functional hypotheses was a detailed knowledge of what the artifacts were used for and how. In 1964 *Prehistoric Technology,* a summary of the studies of tool function conducted by the Russian prehistorian S. A. Semenov, was published in an English translation. Semenov and his colleagues at the Leningrad Academy of Sciences had established the fact that tools of even the hardest stone retained actual traces of their use in the form of polishes, striations and other alterations of the tools' working edges. More often than not the traces of wear were visible only at quite high magnifications. It seemed to scholars in Britain and America that at last the means were in hand for pursuing just the kind of information about tool function that the new hypotheses required.

Semenov's functional interpretations of the uses of Paleolithic and later stone implements unearthed in the U.S.S.R. were fascinating but also tantalizing. He had not included a detailed account of the methodology that formed the basis for his interpretations. To make matters worse, the particular kinds of microscopic equipment employed by Semenov were then available only in the U.S.S.R., and so the translator had omitted most of the few technical details Semenov had included in his original.

As a result a number of prehistorians outside the U.S.S.R. proceeded to do microwear analysis armed only with the translation of Semenov's book and stereoscopic microscopes that often had a maximum magnification of 80 diameters. In addition to this technical handicap the implements these workers selected for study were made from stone materials quite unlike those found in the U.S.S.R. Disappointment and disillusionment followed as one investigator after another found Semenov's results impossible to substantiate.

This situation, however, was scarcely surprising. For one thing, in most cases the investigators could not even see microwear features such as the polishes and striations Semenov had observed because the magnifications they were working with were far too low. For another, the low-magnification wear features they could see (primarily edge damage, the small breaks and flake scars on the working edge of the tool) did not allow precise and unambiguous interpretations of tool function. Many investigators came to the conclusion that Semenov's interpretations were suspect

ALTERED MICROTOPOGRAPHY of the working surfaces of flint tools is seen in the scanning electron micrographs on the opposite page. At the top left is the edge of an unused flint; it is magnified 300 diameters. At the top right is a closer view of an unused flint surface, magnified 1,700 diameters. The author used the flint edge and surface in the middle (magnified 140 and 1,700 diameters respectively) to scrape dry hide. The tool edge is markedly rounded, and the topography of the tool surface has been altered by contact with the hide so as to acquire an extreme matte texture. The author used the edge and surface seen at the bottom (magnified 70 and 130 diameters respectively) to scrape bone. Both edge and surface show the uneven topography produced by such work; characteristic small pits have developed on the flint surface.

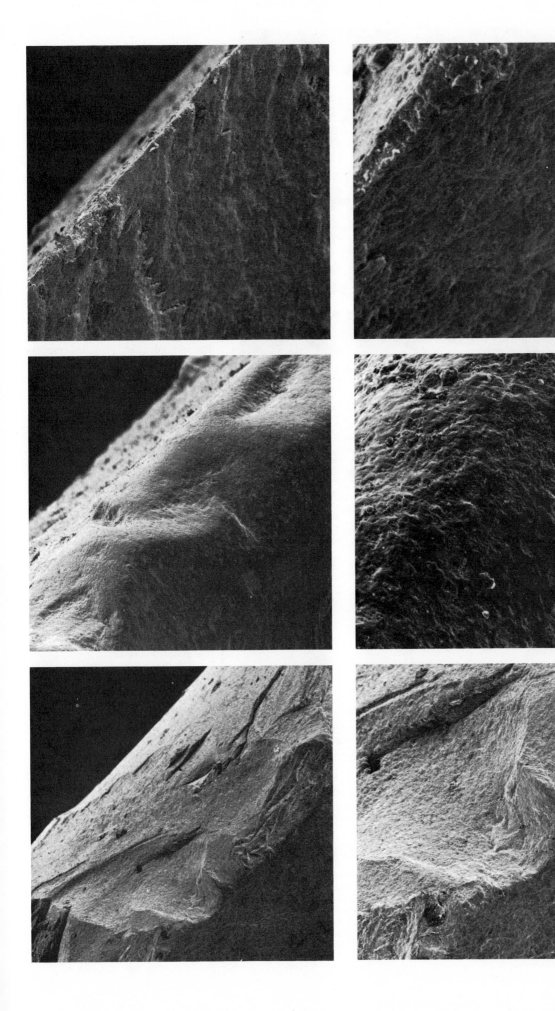

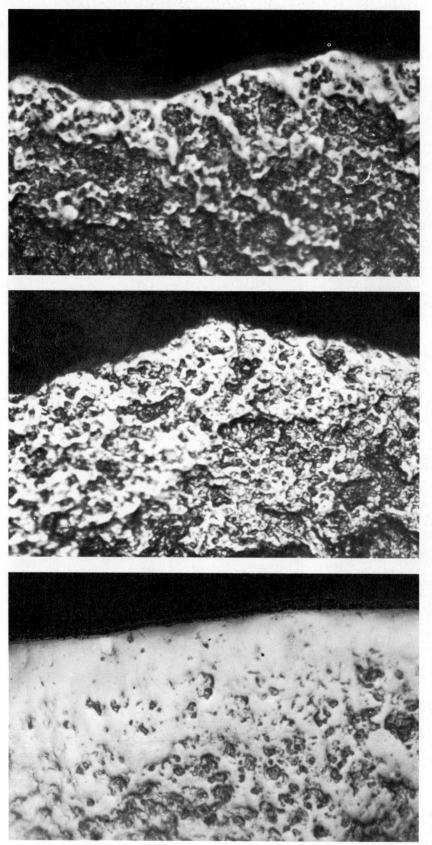

WOOD POLISHES produced on the working edge of three modern replicas of flint tools are visible in these micrographs; all enlarge the surface 300 diameters. The author scraped yew wood with the tool at the top. A characteristically bright wood polish has begun to appear on the elevated parts of the flint. He whittled birch wood with the tool in the middle; the extent of the polish is attributable in part to the wider contact between tool and workpiece. He scraped yew extensively with the tool at the bottom; the depressions on its edge are almost obliterated.

and that microwear analysis simply did not work. Nevertheless, the demand for information about the functions of stone tools ensured that the research would continue.

As a result of this chain of events investigators outside the U.S.S.R. concentrated on studies of the edge damage that could be observed with low-powered stereomicroscopes and ignored the polishes and striations that only begin to be visible at a magnification of 200 diameters. Many edge-wear studies sensibly relied on experiments. Modern replicas of Paleolithic implements were made and were used in various ways to work on a wide range of materials in order to determine whether the resulting traces of wear differed from material to material. Most of these programs, however, involved too few experiments, controlled too few variables and were too limited in scope to achieve anything useful.

The one adequate program employing the low-magnification approach to the analysis of edge damage was conducted at Harvard University by Ruth Tringham and her students. When the results of the work were published in 1974, the chief demonstrable distinction proved to be one between work on "hard" materials (such as bone, antler and wood) and work on "soft" materials (such as meat, hides and nonwoody plant materials). No reliable criteria were found for distinguishing between different methods of working, such as scraping, whittling, sawing, cutting and the like. It was also impossible to distinguish between on the one hand edge-damage scars resulting from the actual use of an implement and on the other hand small scars created in the course of manufacture or by the implement's rubbing against other hard materials during millenniums of burial.

I first undertook research in microwear in 1972 after a review of the literature in the field and some preliminary studies. These preliminaries convinced me that I should employ a wider range of microscope magnifications and techniques than others had. I began with a program of experiments designed to provide a framework for analysis of the functions served by particular sets of flint implements from English sites of the Lower Paleolithic: 500,000 to 100,000 years ago.

I had three microscopes at my disposal: a light stereomicroscope with a range of magnifications between six and 50 diameters, a light microscope with a range between 50 and 1,000 diameters, and my principal research instrument, a microscope with an incident-light attachment and a range between 24 and 400 diameters. I also made occasional use of a scanning electron microscope, mainly for magnifications above 500 diameters.

After making replicas of Paleolithic stone implements I conducted a series of nearly 200 tests, processing a variety of foodstuffs and other materials in many different ways. I also subjected certain implements to the kinds of natural wear that are likely either to make microscopic scars similar to those made by human use or to erase such scars. Along this same line I was able, thanks to the availability of large numbers of Paleolithic implements that had been subjected to wear by soil movements, chemical weathering and abrasion by waterborne and wind-borne sediments, to compare this natural kind of wear with my experimental results.

The key finding that emerged from those tests was that microwear polishes on the working edges of modern replicas become visible at magnifications between 100 and 400 diameters under illumination striking the sample at an angle of 90 degrees to the optical axis of the microscope. The different kinds of polish can readily be distinguished from one another. Whether the activity was cutting or whittling wood, cutting bone, cutting meat or scraping skins, I found that each produced a characteristic kind of work polish.

The work polishes proved to be durable; they could not be removed from my replica implements even with chemical cleaning. I applied caustics that ran the full pH spectrum from an extreme base (sodium hydroxide) to an extreme acid (hydrochloric) without effect. The same was true with various organic solvents. I concluded that the work polishes represent real and permanent alterations in the microtopography of the flint. Accordingly similar polishes seemed likely to have survived unaltered on flint artifacts of great age. This being the case, it should be possible to infer from the traces of microwear observable on a Pa-

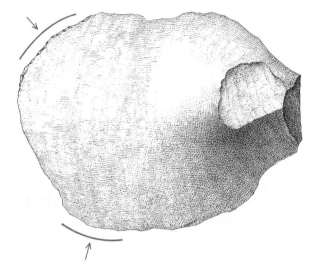

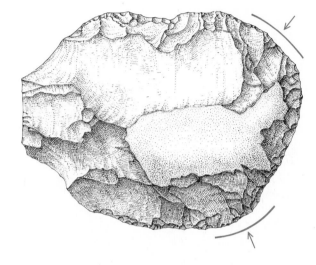

PALEOLITHIC FLAKE TOOL, a "side scraper" from Hoxne, an Acheulean site in England, was among some 800 flint implements examined for evidences of microwear by the author. The top of the flake (*right*) still shows some of the outer surface (*lighter area*) of the nodule of flint the flake was struck from. Lines and arrows (*color*) indicate the working edges of the tool. Prehistorians have assumed that scrapers were used to process animal hides. Microwear traces found on the flake lend support to such an assumption (*see illustration below*).

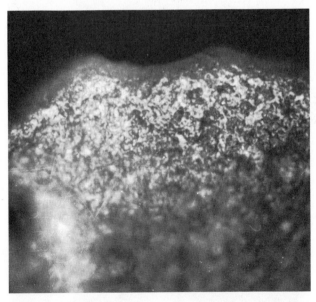

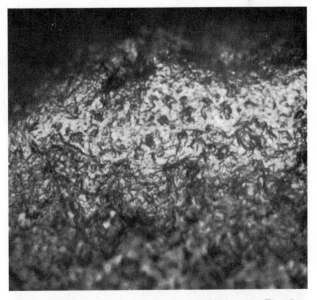

REPLICA AND ORIGINAL are compared in these micrographs. The tool edge at the left was used by the author to scrape dry pigskin for an hour. The edge, seen here magnified 300 diameters, developed a dull matte work polish and had become rounded by wear. The edge of Hoxne scraper (*right*) is seen magnified 300 diameters. It shows the same dull matte polish and wear-rounding the replica tool does.

leolithic tool just what use that particular tool had served.

The distinctive microwear polishes can be described as follows.

Wood polish: The tool edge shows a polish that is consistent in appearance regardless of whether the wood being worked is hard, soft, fresh or seasoned. The polish is also the same regardless of the manner of tool use. It is very "bright," reflecting a considerable percentage of the incident illumination, and very smooth in texture. Because the polish first develops on the elevated parts of the microtopographic surface of the flint its gross appearance is affected by that topography up to the point where the contact area becomes completely polished. Thus if the original topography of the flint is coarse, the polish in its initial stages will be distributed in a net-like pattern. If the flint is fine-grained, the polish is soon evenly spread. Regardless of the distribution, the polish has a constant bright, smooth character.

Bone polish: The tool edge is bright, but the polish has a rough, uneven tex-

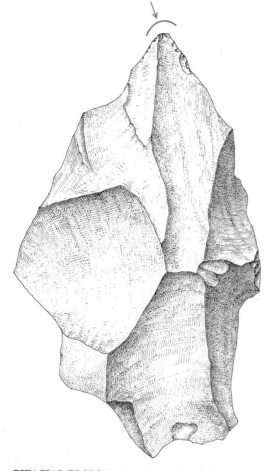

BIFACIAL TOOLS from two Lower Paleolithic sites in England are a "chopper" from Clacton-on-Sea (*left*) and a "hand axe" from the Acheulean site Hoxne. Lines and arrows (*color*) locate their working edges. The microwear analysis shows they served different functions.

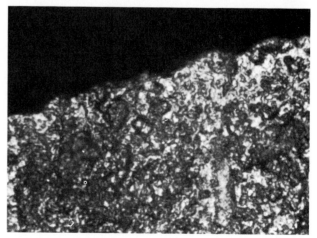

TWO WORKING EDGES are magnified 300 diameters. At the left is one edge of the "bit" of the Clacton biface. The presence of wood polish indicates that the supposed "chopper" was used for woodwork- ing; damage to the bit point, not visible here, indicates that the tool was used to bore holes. The edge of the Hoxne "axe" at right shows dull, greasy work polish characteristic of meat-cutting implements.

ture that lacks the smoothness characteristic of wood polish. One distinctive feature of the rough texture of bone polish is the presence of numerous pits on the otherwise bright surface. Bone polish develops more slowly than wood polish. On a modern replica, even after prolonged use, the polish is seldom very extensively developed. My experiments revealed no consistent differences between the polishes on tools used to work cooked bone and those on tools for working uncooked bone, or between the polishes on tools used to work bone belonging to different species of animals.

Hide polishes: Here the tool edges do not develop a single distinctive kind of polish. The hide polishes differ depending on the material being worked. They range from a relatively bright polish with a greasy appearance (produced by working fresh wet hide) to a dull matte polish (produced by working dry hide or leather). The differences are attributable to variations in the quantity of lubricants present in the animal skin at different stages. A fresh hide gradually creates a polish not unlike that created by the cutting of meat. As the hide becomes progressively drier it contains progressively less lubricant, and the tool polish not only develops faster but also is duller and less greasy in appearance. If the hide is fully dried or tanned, the polish is quite dull and shows an extreme matte texture. Regardless of these differences in polish all hide-working tools show two characteristic kinds of microwear. One is relatively severe attrition of the working edge of the implement, that is, removal of flint by means other than breakage or scratching. This attrition gives the stone implement a markedly rounded edge. The other characteristic is the development of shallow and diffuse linear surface features that run parallel to the direction in which the tool is moved. These diffuse linear marks are similar to the striations caused by other kinds of materials, but they cannot be mistaken for such striations, which are much more prominent.

Meat polish: The tool edge that is used to slice meat and other soft animal tissue develops a microwear polish rather like the polish produced by working fresh hide. This polish is easily distinguished, however, from the polishes created by the working of dry hide, bone, antler, wood and nonwoody plant materials. Pronouncedly greasy, it is at the same time dull rather than bright. Thus with respect to brightness the contrast between meat polish and an unaltered flint surface is slight. For this reason meat polish does not show up well in photomicrography. The distinction is nonetheless clear to the eye. The grainy texture characteristic of raw flint is replaced by a matte texture that, although it seems to preserve the original surface microtopography, has actually transformed the elevations and depressions into a semicontinuous surface. Tools that show meat polish also frequently bear short, narrow striations.

Antler polishes: The edges of tools used to work antler exhibit one or another of two distinctive polishes. The difference depends on how the tool was used. Scraping, planing or graving antler leaves a very bright and smooth polish. Sawing antler, however, leaves a polish like bone polish: it is bright but pitted. In its early stages of development smooth antler polish is sometimes virtually indistinguishable from wood polish. When it is further developed, the polished surface displays small scattered depressions, giving it a pockmarked appearance that is quite different both from a wood polish and from the stronger surface pitting characteristic of the rougher antler polish. My experiments with antler were conducted almost entirely with samples that had been soaked for a day or two in water. Dry antler is so hard that stone tools used to work it are dulled by edge damage before anything has been accomplished. Watersoaked antler, however, is quite easy to work.

Nonwoody plant polishes: The edges of tools used to cut nonwoody plant stems, such as grasses or bracken, acquire a "corn gloss." The characteristic feature is a very smooth, highly reflective surface with a "fluid" appearance. If any striations are present, they often appear to be "filled in." At the same time the polished surfaces of the working edge develop curious comet-shaped pits. As the term implies, corn gloss is most commonly found on the flint sickles used by Neolithic farmers to harvest domesticated species of the grass family. As I was to discover, however, some nonwoody plants were cut in Lower Paleolithic times, and the cutting tools developed the same kind of gloss.

Work polishes alone enable the investigator to infer what materials were processed with various flint implements.

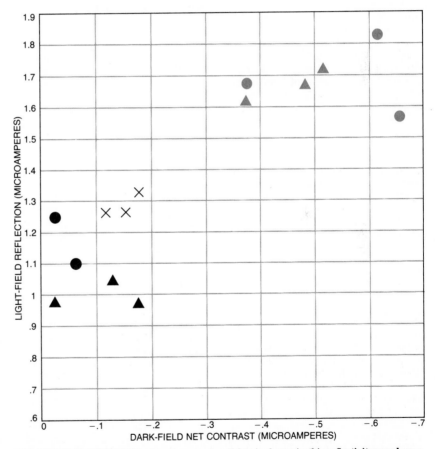

RELATIVE BRIGHTNESS of various work polishes is shown in this reflectivity graph as a function of two measurements. The ordinate values indicate the amount of light reflected from a standard area of polished surface under normal light-field illumination, as registered in microamperes on a photometer. The abscissa values indicate differences between the reflectivity of polished and unpolished surfaces of an implement under dark-field illumination: the smaller the difference, the rougher the texture of the polished surface. The brightest and smoothest of the work polishes were the "corn glosses" produced by cutting nonwoody plant stems rich in plant opal. Colored dots indicate the readings on Neolithic flint sickle blades from Syria; the triangles, the readings on Neolithic blades from Bavaria. Among the dullest and roughest of the work polishes were the ones formed on modern replicas by work on greasy hide (black triangles); the polishes produced by work on dry hides (black dots) were rougher but brighter. Polishes produced by working wood (black crosses) were smoother and brighter.

If one is to determine how the implements were used, however, one must rely on several other kinds of microwear evidence. Perhaps most important are the distribution and orientation of such linear wear features as striations. Other kinds of evidence include the location and nature of edge damage and the location and extent of the polished working portions. All such evidence must be considered in relation to the general size and shape of the tool. In the broadest terms, once an inventory of the various kinds of microwear evident on a particular implement has been made, one then asks how the tool must have been handled to acquire the observed features. For example, microwear traces on both sides of a working edge, combined with striations that run parallel to the edge, are the strongest kind of evidence that the implement was used for sawing or cutting. Analysis of the work polish should then indicate what material was sawed or cut.

Having established six broad categories of polishes, I was prepared to apply my experimental results to selected Paleolithic artifacts. A skeptical colleague suggested, however, that I first submit my analytical technique to a blind test. The colleague, Mark H. Newcomer of the University of London, had strong doubts about the validity of microwear analysis. We agreed that he would make several replicas of flint tools and then work on various materials with them. After recording what he had done with the tools and then cleaning them, he would send them to me for analysis. Thereafter we would meet and compare my inferences with his records of the actual uses. Newcomer made 15 replicas of ancient flint tools and did various kinds of work with a total of 16 tool edges.

The results of the blind test were instructive. To be sure, the number of implements was small. Nevertheless, I identified the working portions of the tool edge in 14 edges of the 16. For 12 of the edges I was able to reconstruct the mode of tool use and for 10 of them to infer the kind of material worked.

Some of the inferences were remarkably close to the mark. For example, Newcomer had skinned a hare with a double-edged tool, using one of the edges for the actual skinning and the opposite edge to sever those parts of the limbs that remained with the skin during hide preparation. I identified the wear on the skinning edge as meat-cutting polish. (I had no way of knowing that in this instance the meat was less than a millimeter below the skin.) The microwear on the opposite edge I interpreted as the result of breaking a joint.

With another implement Newcomer had cut fresh meat resting on a wood cutting board. I was able to distinguish between the microwear caused by the cutting of the meat and the incidental wear caused by the contact between the flint and the cutting board.

Even some of my misinterpretations were not unreasonable. For example, Newcomer had used the edge of one flint tool to cut frozen meat, which leaves few traces of wear. He had cut the meat on a wood board, however, and contact with the board did leave discernible traces. I interpreted the resulting microwear as characteristic of an implement used very delicately on wood. Since Newcomer's tests were the first check on the validity of high-resolution microwear analysis, I found the results quite encouraging.

It was now time to apply the technique to selected Paleolithic artifacts. Three classic British sites of the Lower Paleolithic period met the desired criteria. First, flint implements from all three sites are well preserved; they have not accumulated the surface patina that would conceal or destroy the evidence of microwear, and they have usually escaped damaging natural abrasion. Second, all the artifacts had been recently excavated, ensuring that their stratigraphic position in the ground had been recorded under strict controls and that they had been carefully handled and stored to eliminate the danger of post-excavation damage. The sites were at Clacton-on-Sea in Essex (the "Golf Course site"), at Swanscombe in Kent (the "Lower Loam") and at Hoxne in Suffolk (mainly the "Lower Industry").

The Clacton site has given the name Clactonian to an entire Lower Paleolithic flint-tool industry that flourished some 250,000 years ago during the early stages of the Mindel-Riss interglacial period. The distinctive flake tools of the Clactonian industry were made by striking rather coarse flakes off nodules of flint and trimming a few of the flakes into the desired shape. Some of the leftover "cores" of flint were also employed as tools.

The Lower Loam at Swanscombe is a somewhat later Clactonian site, occupied during the same Mindel-Riss interglacial period. The stone artifacts from the Lower Loam include tools made on flakes. Hoxne, a still later site, has yielded refined tools worked on both sides. These "bifacial" implements are typologically assigned to the Acheulean, a Lower Paleolithic industry named after Saint-Acheul, the site in France where such implements were first found. The Hoxne strata also contain an abundance of flint flakes, many of them the waste left over from production of the bifacial tools.

The total number of artifacts in suitable condition for microwear analysis was not large. The Clacton group included 144 tools from a layer of gravel

at the Golf Course site and 102 from a layer of marl. Some of the flakes could be fitted into the original core from which they had been struck, indicating that they had been made on the spot. Taken together with microwear evidence that the flake tools had been used for butchering, woodworking, hide-working and some work on bone, this suggests that the Golf Course site was probably occupied for some time rather than being a transient hunters' camp. The predominant activities at the site were woodworking and butchering.

Of the artifacts from Clacton that I examined 22 were the coarse bifacial tools that are traditionally classified as choppers. Microwear indicates that only two of the 22 had actually been used as tools. This is a utilization rate of 10 percent, about half the rate for the flakes found at the site. Of the flakes from the gravel 22 percent showed traces of use; of those from the marl 16 percent did. The relative ratios suggest that the Clacton toolmakers were primarily interested in using their flint cores to turn out flakes, as opposed to bifaces.

Sixty-six flake tools from the Lower Loam at Swanscombe were in suitable condition for microwear analysis. Of these only four actually showed traces of use. The microwear characteristics shown by the four flakes were much like those visible on flake tools from Clacton. The Swanscombe sample is too small, however, to allow any conclusion from this coincidence.

The artifacts from Hoxne included one group of tools (Lower Industry, Layer 3 West) with little or no abrasion damage. I studied the entire assemblage from that layer, numbering 408 implements. I also analyzed a random sample of artifacts from other Lower Industry and Upper Industry strata. The Acheulean industry at Hoxne, with its emphasis on the manufacture of bifacial tools, is marked by large quantities of flakes that must be counted as potential implements even though most of them are surely the debris of toolmaking, too thin-edged and fragile to be made into flake tools. Indeed, microwear analysis reveals that only 9 percent of the flakes from all the Lower Industry levels actually show evidence of wear.

The makers of the Hoxne tools used them for butchering, woodworking, hide-working and for boring wood and bone. Interestingly enough, some were also used to slice or cut plant material other than wood. These hunters may have gathered reeds or bracken for bedding. The butchering was not done exclusively with flake tools: two of the Lower Industry "hand axes" showed the polish characteristic of butchering implements.

Among the Upper Industry tools at

Hoxne were a small number of the flake implements that are traditionally called "side scrapers" and are presumed to have played a role in the dressing of hides. The microwear on these tools lends support to the guess of the traditionalists; most of the side scrapers show the polish characteristic of hide-working tools.

To cite one further example of microwear analysis, a bifacial tool from Clacton was found to show wood polish on its working surface. Further examination revealed utilization damage that could only have come from a rotary motion such as boring; the tool had been turned in a clockwise direction at the same time that downward pressure was being applied. Similar wear patterns also appear on flake tools that were used for boring. The patterns suggest that the Clacton woodworkers of perhaps 200,000 years ago were consistently right-handed.

The seeming wastefulness represented by the 9 percent rate of flake utilization at Hoxne may be more apparent than real; most of the flakes were biface-manufacture waste and unsuitable for use as tools. The prodigal use of flint at Clacton cannot be similarly explained away as the debris of bifacial-tool production. Perhaps at both sites much of the waste is better explained by the fact that chalk flint, an excellent raw material for the making of stone tools, can be found easily almost anywhere in southeastern England: in river gravels, on beaches and other superficial deposits and of course in exposures of the chalk itself.

The microwear analysis of work polishes on this group of Lower Paleolithic implements provides the first direct and unequivocal evidence of the kinds of human activity that took place at English campsites roughly 250,000 years ago. Such findings make it clear that a new and rewarding method of archaeological research has finally come of age. It is now possible, assuming that the tools have been suitably preserved, to determine in most instances not only how ancient flint tools were used but also what they were used on.

It seems very likely, although it remains to be proved, that microwear that can be interpreted in similar ways is present on tools made from stone materials other than flint, such as obsidian, chert and even fine-grained basalts and quartzites. I have found this to be true of one fine-grained chert from southern Africa; experiments show that the material retains microwear polishes that are directly comparable to those found on chalk flint. The information derived from future microwear studies should enable prehistorians to discuss with increasing confidence the technology and economy of early man.

11

The Food-Sharing Behavior of Protohuman Hominids

by Glynn Isaac
April 1978

*Excavations at two-million-year-old sites in East Africa
offer new insights into human evolutionary progress
by showing that early erect-standing hominids made
tools and carried food to a home base*

Over the past decade investigators of fossil man have discovered the remains of many ancient protohumans in East Africa. Findings at Olduvai, Laetolil, Koobi Fora, the Omo Valley and Hadar, to name some prominent locations, make it clear that between two and three million years ago a number of two-legged hominids, essentially human in form, inhabited this part of Africa. The paleontologists who have unearthed the fossils report that they differ from modern mankind primarily in being small, in having relatively large jaws and teeth and in having brains that, although they are larger than those of apes of comparable body size, are rarely more than half the size of modern man's.

The African discoveries have many implications for the student of human evolution. For example, one wonders to what extent the advanced hominids of two million years ago were "human" in their behavior. Which of modern man's special capabilities did they share? What pressures of natural selection, in the time since they lived, led to the evolutionary elaboration of man's mind and culture? These are questions that paleontologists find difficult to answer because the evidence that bears on them is not anatomical. Archaeologists, by virtue of their experience in studying prehistoric behavior patterns in general, can help to supply the answers.

It has long been realized that the human species is set apart from its closest living primate relatives far more by differences in behavior than by differences in anatomy. Paradoxically, however, the study of human evolution has traditionally been dominated by work on the skeletal and comparative anatomy of fossil primates. Several new research movements in recent years, however, have begun to broaden the scope of direct evolutionary inquiry. One such movement involves investigations of the behavior and ecology of living primates and of other mammals. The results of these observations can now be compared with quantitative data from another new area of study, namely the cultural ecology of human societies that support themselves without raising plants or animals: the few surviving hunter-gatherers of today. Another important new movement has involved the direct study of the ecological circumstances surrounding human evolutionary developments. Investigations of this kind have become possible because the stratified sedimentary rocks of East Africa preserve, in addition to fossil hominid remains, an invaluable store of data: a coherent, ordered record of the environments inhabited by these protohumans.

The work of the archaeologist in drawing inferences from such data is made possible by the fact that at a certain stage in evolution the ancestors of modern man became makers and users of equipment. Among other things, they shaped, used and discarded numerous stone tools. These virtually indestructible artifacts form a kind of fossil record of aspects of behavior, a record that is complementary to the anatomical record provided by the fossil bones of the toolmakers themselves. Students of the Old Stone Age once concentrated almost exclusively on what could be learned from the form of such tools. Today the emphasis in archaeology is increasingly on the context of the artifacts: for example the distribution pattern of the discarded tools in different settings and the association of tools with various kinds of food refuse. A study of the contexts of the early African artifacts yields unique clues both to the ecological circumstances of the protohuman toolmakers and to aspects of their socioeconomic organization.

Comparing Men and Apes

What are the patterns of behavior that set the species *Homo sapiens* apart from its closest living primate relatives? It is not hard to draw up a list of such differences by comparing human and ape behavior and focusing attention not on the many features the two have in common but on the contrasting features. In the list that follows I have drawn on recent field studies of the great apes (particularly the chimpanzee, *Pan troglodytes*) and on similar studies of the organization of living hunter-gatherer societies. The list tends to emphasize the contrasts relating to the primary subsistence adaptation, that is, the quest for food.

First, *Homo sapiens* is a two-legged primate who in moving from place to place habitually carries tools, food and other possessions either with his arms or in containers. This is not true of the great apes with regard to either posture or possessions.

Second, members of *Homo sapiens* so-

PAST AND PRESENT LANDSCAPES in the Rift Valley region of East Africa, shown schematically on the opposite page, summarize the geological activity that first preserved and later exposed evidence of protohuman life. Two million years ago (*top*) the bones of hominids (*1–4, color*) and other animals (*x's, color*) were distributed across hills and a floodplain (*foreground*) adjacent to a Rift Valley lake. Also lying on the surface were stone tools (*black dots*) made, used and discarded by the protohumans. Layers of sediments then covered the bones and tools lying on the floodplain; burial preserved them, whereas the bones and tools in the hills were eventually washed away. Today (*bottom*), after a fault has raised a block of sediments, erosion is exposing some of the long-buried bones and clusters of tools, including the three types of site shown on the surface in the top block diagram (*A–C*). Sites of Type A contain clusters of stone tools together with the leftover stone cores that provided the raw material for the tools and waste flakes from the toolmaking process, but little or no bone is present. Sites of Type B contain similar clusters of tools in association with the bones of a single large animal. Sites of Type C also contain similar clusters of tools, but the bones are from many different animal species.

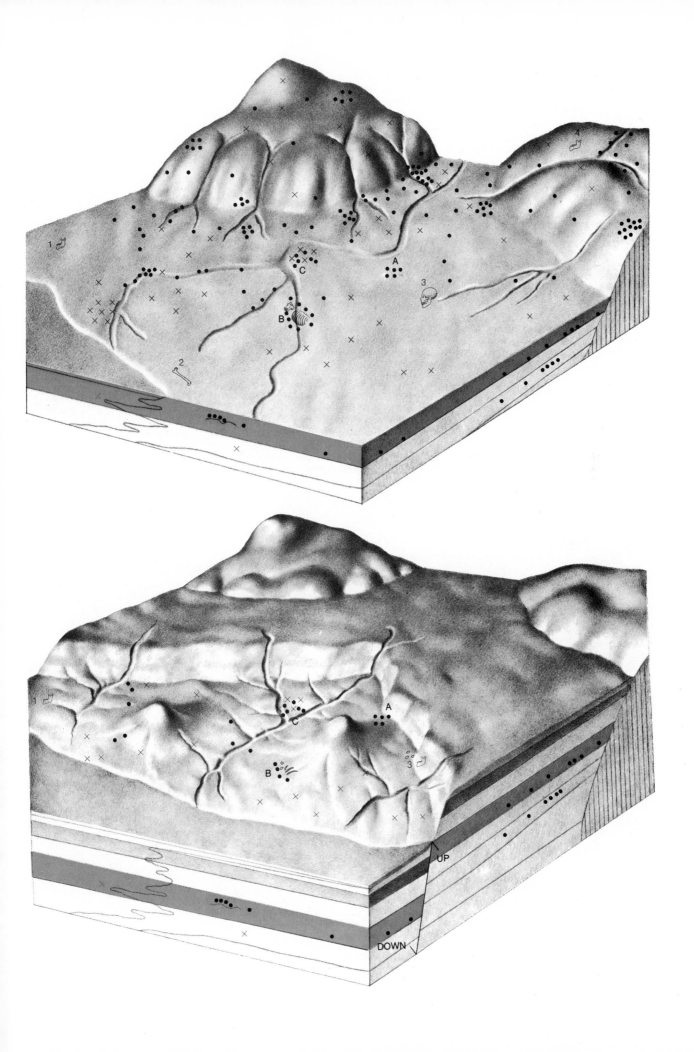

cieties communicate by means of spoken language; such verbal communication serves for the exchange of information about the past and the future and also for the regulation of many aspects of social relations. Apes communicate but they do not have language.

Third, in *Homo sapiens* societies the acquisition of food is a corporate responsibility, at least in part. Among members of human social groupings of various sizes the active sharing of food is a characteristic form of behavior; most commonly family groups are the crucial nodes in a network of food exchange. Food is exchanged between adults, and it is shared between adults and juveniles. The only similar behavior observed among the great apes is seen when chimpanzees occasionally feed on meat. The chimpanzees' behavior, however, falls far short of active sharing; I suggest it might better be termed tolerated scrounging. Vegetable foods, which are the great apes' principal diet, are not shared and are almost invariably consumed by each individual on the spot.

Fourth, in human social groupings there exists at any given time what can be called a focus in space, or "home base," such that individuals can move independently over the surrounding terrain and yet join up again. No such

home base is evident in the social arrangements of the great apes.

Fifth, human hunter-gatherers tend to devote more time than other living primates to the acquisition of high-protein foodstuffs by hunting or fishing for animal prey. It should be noted that the distinction is one not of kind but of degree. Mounting evidence of predatory behavior among great apes and monkeys suggests that the principal contrast between human beings and other living primates with respect to predation is that only human beings habitually feed on prey weighing more than about 15 kilograms.

The gathering activities of human hunter-gatherers include the collection of edible plants and small items of animal food (for example lizards, turtles, frogs, nestling birds and eggs). Characteristically a proportion of these foodstuffs is not consumed until the return to the home base. This behavior is in marked contrast to what is observed among foraging great apes, which almost invariably feed at the spot where the food is acquired.

Still another contrast with great-ape feeding behavior is human hunter-gatherers' practice of subjecting many foodstuffs to preparation for consumption, by crushing, grinding, cutting and heat-

ing. Such practices are not observed among the great apes.

Human hunter-gatherers also make use of various kinds of equipment in the quest for food. The human society with perhaps the simplest equipment ever observed was the aboriginal society of Tasmania, a population of hunter-gatherers that was exterminated in the 19th century. The inventory of the Tasmanians' equipment included wood clubs, spears and digging sticks, cutting tools made of chipped stone that were used to shape the wood objects, and a variety of containers: trays, baskets and bags. The Tasmanians also had fire. Although such equipment is simple by our standards, it is far more complex than the kind of rudimentary tools that we now know living chimpanzees may collect and use in the wild, for example twigs and grass stems.

In addition to this lengthy list of subsistence-related behavioral contrasts between human hunter-gatherers and living primates there is an entire realm of other contrasts with respect to social organization. Although these important additional features fall largely outside the range of evidence to be considered here, they are vital in defining human patterns of behavior. Among them is the propensity for the formation of long-

DESOLATE LANDSCAPE in the arid Koobi Fora district of Kenya is typical of the kind of eroded terrain where gullying exposes both bones and stone tools that were buried beneath sediments and volcanic ash more than a million years ago. Excavation in progress (*center*) **is exposing the hippopotamus bones and clusters of artifacts that had been partially bared by recent erosion and were found by Richard Leakey in 1969. The site is typical of the kind that includes the remains of a single animal and many tools manufactured on the spot.**

term mating bonds between a male and one or more females. The bonds we call "marriage" involve reciprocal economic ties, joint responsibility for aspects of child-rearing and restrictions on sexual access. Another such social contrast is evident in the distinctively human propensity to categorize fellow members of a group according to kinship and metaphors of kinship. Human beings regulate many social relations, mating included, according to complex rules involving kinship categories. Perhaps family ties of a kind exist among apes, but explicit categories and rules do not. These differences are emphasized by the virtual absence from observed ape behavior of those distinctively human activities that are categorized somewhat vaguely as "symbolic" and "ritual."

Listing the contrasts between human and nonhuman subsistence strategies is inevitably an exercise in oversimplification. As has been shown by contemporary field studies of various great apes and of human beings who, like the San (formerly miscalled Bushmen) of the Kalahari Desert, still support themselves without farming, there is a far greater degree of similarity between the two subsistence strategies than had previously been recognized. For example, with regard to the behavioral repertories involving meat-eating and tool-using the differences between ape and man are differences of degree rather than of kind. Some scholars have even used the data to deny the existence of any fundamental differences between the human strategies and the nonhuman ones.

It is my view that significant differences remain. Let me cite what seem to me to be the two most important. First, whereas humans may feed as they forage just as apes do, apes do not regularly postpone food-consumption until they have returned to a home base, as human beings do. Second, human beings actively share some of the food they acquire. Apes do not, even though chimpanzees of the Gombe National Park in Tanzania have been observed to tolerate scrounging when meat is available.

From Hominid to Human

Two complementary puzzles face anyone who undertakes to examine the question of human origins. The first relates to evolutionary divergence. When did the primate stock ancestral to the living apes diverge from the stock ancestral to man? What were the circumstances of the divergence? Over what geographical range did it take place? It is not yet established beyond doubt whether the divergence occurred a mere five to six million years ago, as Vincent M. Sarich of the University of California at Berkeley and others argue on biochemical grounds, or 15 to 20 million years ago, as many paleontologists believe on the grounds of fossil evidence.

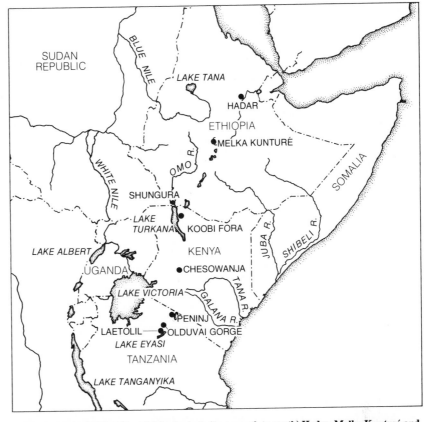

PROMINENT SITES in East Africa include (from north to south) Hadar, Melka Kunturé and Shungura in Ethiopia, the Koobi Fora district to the east of Lake Turkana in Kenya, Chesowanja in Kenya and Peninj, Olduvai Gorge and Laetolil in Tanzania. Dates for clusters of stone tools, some associated with animal bones, uncovered at these sites range from one million years ago (Olduvai Upper Bed II) to 2.5 million (Hadar upper beds). Some sites may be even older.

	OLDUVAI	KOOBI FORA	OMO VALLEY	OTHER
1.0				
1.2	UPPER BED II			PENINJ
	MIDDLE BED II	KARARI SITES		MELKA KUNTURÉ
1.4				CHESOWANJA
1.6	LOWER BED II			
	BED I	KBS, HAS		
1.8				
2.0				
2.2			SHUNGURA MEMBER F	
			SHUNGURA MEMBER E	
2.4				
		?		HADAR UPPER BEDS
2.6				
2.8				
3.0				
3.2				HADAR LOWER BEDS
3.4				
				LAETOLIL
3.6				

YEARS BEFORE PRESENT (MILLIONS)

RELATIVE ANTIQUITY of selected sites in East Africa is indicated in this table. Olduvai Gorge beds I and II range from 1.8 to 1.0 million years in age. The Shungura sites in the Omo Valley are more than two million years old. Two Koobi Fora locales, the hippopotamus/artifact site (HAS) and the Kay Behrensmeyer site (KBS), are at least 1.6 million years old. Initial geological studies of the Koobi Fora sites suggested that they might be 2.5 million years old (colored line). Only hominid fossils have been found in the lower beds at Hadar and at Laetolil.

At least one fact is clear. The divergence took place long before the period when the oldest archaeological remains thus far discovered first appear. Archaeology, at least for the present, can make no contribution toward solving the puzzle of the split between ancestral ape and ancestral man.

As for the second puzzle, fossil evidence from East Africa shows that the divergence, regardless of when it took place, had given rise two to three million years ago to populations of smallish two-legged hominids. The puzzle is how to identify the patterns of natural selection that transformed these protohumans into humans. Archaeology has a major contribution to make in elucidating the second puzzle. Excavation of these protohuman sites has revealed evidence suggesting that two million years ago some elements that now distinguish man from apes were already part of a novel adaptive strategy. The indications are that a particularly important part of that strategy was food-sharing.

The archaeological research that has inspired the formulation of new hypotheses concerning human evolution began nearly 20 years ago when Mary Leakey and her husband Louis discovered the fossil skull he named "Zinjanthropus" at Olduvai Gorge in Tanzania. The excavations the Leakeys undertook at the site showed not only that stone tools were present in the same strata that held this fossil and other hominid fossils but also that the discarded artifacts were associated with numerous broken-up animal bones. The Leakeys termed these concentrations of tools and bones "living sites." The work has continued at Olduvai under Mary Leakey's direction, and in 1971 a major monograph was published that has made the Olduvai results available for comparative studies.

Other important opportunities for archaeological research of this kind have come to light in the Gregory Rift Valley, at places such as the Koobi Fora (formerly East Rudolf) region of northern Kenya, at Shungara in the Omo Valley

of southwestern Ethiopia and in the Hadar region of eastern Ethiopia. Current estimates of the age of these sites cover a span of time from about 3.2 million years ago to about 1.2 million.

Since 1970 I have been co-leader with Richard Leakey (the son of Mary and Louis Leakey) of a team working at Koobi Fora, a district that includes the northeastern shore of Lake Turkana (the former Lake Rudolf). Our research on the geology, paleontology and paleoanthropology of the district involves the collaboration of colleagues from the National Museum of Kenya and from many other parts of the world. Work began in 1968 and has had the help and encouragement of the Government of Kenya, the National Science Foundation and the National Geographic Society. Our investigations have yielded archaeological evidence that corroborates and complements the earlier evidence from Olduvai Gorge. The combined data make it possible to see just how helpful archaeology can be in answering

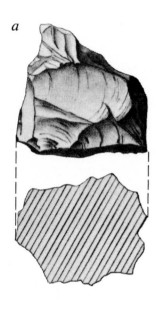

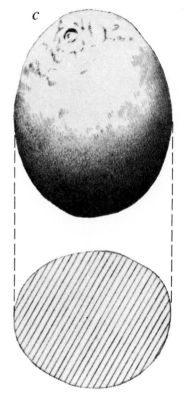

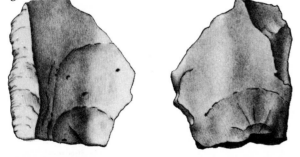

KOOBI FORA ARTIFACTS include four from the HAS assemblage (*left*) and four from the KBS assemblage (*right*). All are shown actual size; the stone is basalt. The HAS core (*a*) shows what is left of a piece of stone after a number of flakes have been struck from it by percussion. The jagged edges produced by flake removal give the core potential usefulness as a tool. The flakes were detached from the core by blows with a hammerstone like the one shown here (*c*). The sharp edges of the flakes, such as the example illustrated (*b*), allow their use as cutting tools. The tiny flake (*d*) is probably an accidental product of the percussion process; the presence of many stone splinters such

questions concerning human evolution.

At Koobi Fora, as at all the other East African sites, deposits of layered sediments, which accumulated long ago in the basins of Rift Valley lakes, are now being eroded by desert rainstorms and transient streams. As the sedimentary beds erode, a sample of the ancient artifacts and fossil bones they contain is exposed at the surface. For a while the exposed material lies on the ground. Eventually, however, the fossil bones are destroyed by weathering or a storm washes away stone and bone alike.

All field reconnaissance in East Africa progresses along essentially similar lines. The field teams search through eroded terrain looking for exposed fossils and artifacts. In places where concentrations of fossil bone or promising archaeological indications appear on the surface the next step is excavation. The digging is done in part to uncover further specimens that are still in place in the layers of sediments and in part to gather exact information about the original stratigraphic location of the surface material. Most important of all, excavation allows the investigators to plot in detail the relative locations of the material that is unearthed. For example, if there are associations among bones and between bones and stones, excavation will reveal these characteristics of the site.

The Types of Sites

The archaeological traces of protohuman life uncovered in this way may exhibit several different configurations. In some ancient layers we have found scatterings of sharp-edged broken stones even though there are no other stones in the sediments. The broken stones come in a range of forms but all are of the kind produced by deliberate percussion, so that we can classify them as undoubted artifacts. Such scatterings of artifacts are often found without bone being present in significant amounts. These I propose to designate sites of Type A.

In some instances a layer of sediment may include both artifacts and animal bones. Such bone-and-artifact occurrences fall into two categories. The first consists of artifacts associated with bones that represent the carcass of a single large animal; these sites are designated Type B. The second consists of artifacts associated with bones representing the remains of several different animal species; these sites are designated Type C.

The discovery of sites with these varied configurations in the sediments at Koobi Fora and Olduvai provides evidence that when the sediments containing them were being deposited some 2.5 to 1.5 million years ago, there was at least one kind of hominid in East Africa that habitually carried objects such as stones from one place to another and made sharp-edged tools by deliberately fracturing the stones it carried with it. How does this archaeological evidence match up with the hominid fossil record? The fossil evidence indicates that

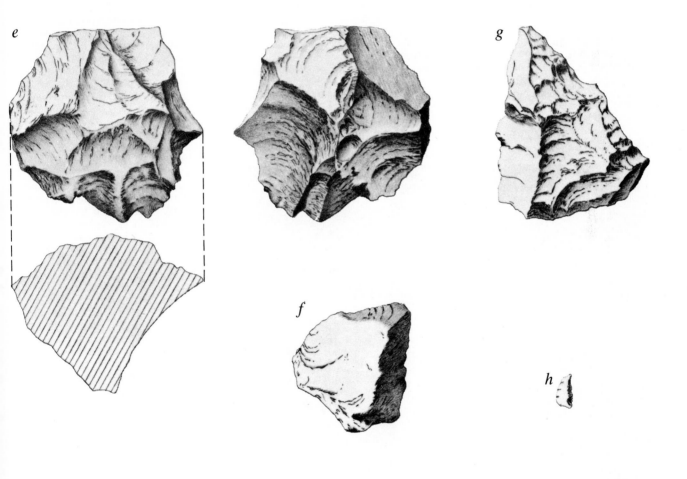

as this one in the HAS tool clusters indicates that the stone tools were made on the spot. At the same time the absence of local unworked stone as potential raw material for tools suggests that the cores were carried to the site by the toolmakers. The artifacts from the second assemblage also include a core (e) that has had many flakes removed by percussion and another small splinter of stone (h). The edges of the two flakes (f, g) are sharp enough to cut meat, hide, sinew or wood. As at the hippopotamus/artifact site, the absence of local raw material for stone tools at the Kay Behrensmeyer site suggests that suitable lumps of lava must have been transported there by the toolmakers.

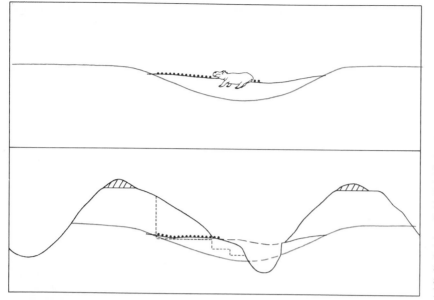

KOOBI FORA LANDSCAPE in the vicinity of the hippopotamus/artifact site consisted of a level floodplain near the margin of a lake (*top section*). Protohuman foragers apparently found the carcass of a hippopotamus lying in a stream-bed hollow and made tools on the spot in order to butcher the carcass. Their actions left a scatter of stone tools among the bones and on the ground nearby. The floodplain was buried under layers of silt and ash and was subsequently eroded (*bottom section*), exposing some bones and tools. Their discovery led to excavation.

two and perhaps three species of bipedal hominids inhabited the area at this time, so that the question arises: Can the species responsible for the archaeological evidence be identified?

For the moment the best working hypothesis seems to be that those hominids that were directly ancestral to modern man were making the stone tools. These are the fossil forms, of early Pleistocene age, classified by most paleontologists as an early species of the genus *Homo*. The question of whether or not contemporaneous hominid species of the genus *Australopithecus* also made tools must be set aside as a challenge to the ingenuity of future investigators. Here I shall simply discuss what we can discover about the activities of early toolmaking hominids without attempting to identify their taxonomic position (or positions).

Reading the Evidence

As examples of the archaeological evidence indicative of early hominid patterns of subsistence and behavior, consider our findings at two Koobi Fora excavations. The first is a locality catalogued as the hippopotamus/artifact site (HAS) because of the presence of fossilized hippopotamus bones and stone tools.

The site is 15 miles east of Lake Turkana. There in 1969 Richard Leakey discovered an erosion gully cutting into an ancient layer of volcanic ash known as the KBS tuff. (KBS stands for Kay Behrensmeyer site; she, the geologist-paleoecologist of our Koobi Fora research team, first identified the ash layer at a nearby outcrop.) The ash layer is the uppermost part of a sedimentary deposit known to geologists as the Lower Member of the Koobi Fora Formation; here the ash had filled in one of the many dry channels of an ancient delta. Leakey found many bones of a single hippopotamus carcass weathering out of the eroded ash surface, and stone artifacts lay among the bones.

J. W. K. Harris, J. Onyango-Abuje and I supervised an excavation that cut into an outcrop where the adjacent delta sediments had not yet been disturbed by erosion. Our digging revealed that the hippopotamus carcass had originally lain in a depression or puddle within an ancient delta channel. Among the hippopotamus bones and in the adjacent stream bank we recovered 119 chipped stones; most of them were small sharp flakes that, when they are held between the thumb and the fingers, make effective cutting implements. We also recovered chunks of stone with scars showing that flakes had been struck from them by percussion. In Paleolithic tool classification these larger stones fall into the category of core tool or chopper. In addition our digging exposed a rounded river pebble that was battered at both ends; evidently it had been used as a hammer to strike flakes from the stone cores.

The sediments where we found these artifacts contain no stones larger than a pea. Thus it seems clear that the makers of the tools had carried the stones here from somewhere else. The association between the patch of artifacts and the hippopotamus bones further suggests that toolmakers came to the site carrying stones and hammered off the small sharp-edged flakes on the spot in order to cut meat from the hippopotamus carcass. We have no way of telling at present whether the toolmakers themselves killed the animal or only came on it

HAMMERSTONE unearthed at the hippopotamus/artifact site is a six-centimeter basalt pebble; it is shown here being lifted from its position on the ancient ground surface adjacent to the hippopotamus bones. Worn smooth by water action before it caught the eye of a toolmaker some 1.7 million years ago, the pebble is battered at both ends as a result of use as a hammer.

dead. Given the low level of stone technology in evidence, I am inclined to suspect scavenging rather than hunting.

The HAS deposit was formed at least 1.6 million years ago. The archaeological evidence demonstrates that the behavior of some hominids at that time differed from the behavior of modern great apes in that these protohumans not only made cutting tools but also ate meat from the carcasses of large animals. The hippopotamus/artifact site thus provides corroboration for evidence of similar behavior just as long ago obtained from Mary Leakey's excavations at Olduvai Gorge.

This finding does not answer all our questions. Were these protohumans roaming the landscape, foraging and hunting, in the way that a troop of baboons does today? Were they instead hunting like a pride of lions? Or did some other behavioral pattern prevail? Excavation of another bone-and-arti-

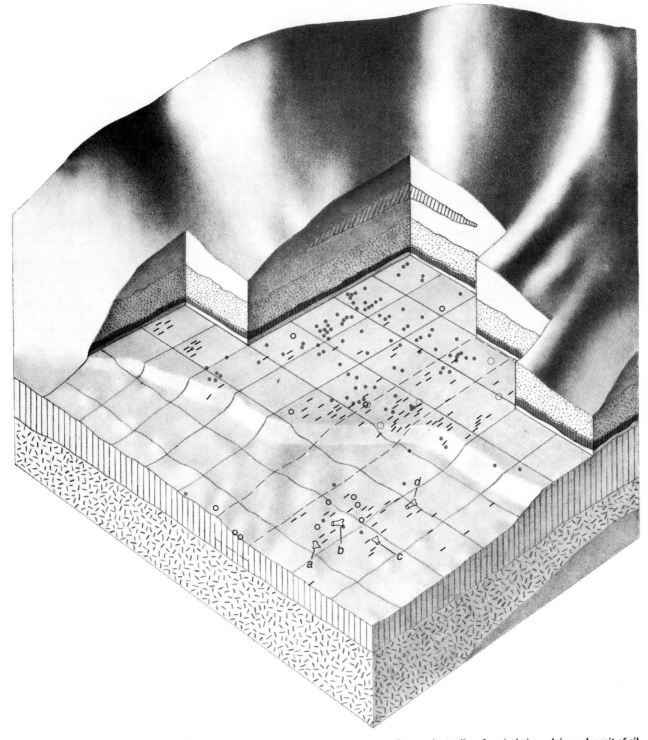

FINDINGS at the hippopotamus/artifact site are shown schematically in this block diagram; squares are one meter to a side. In the foreground are the objects that had been exposed by weathering: hippopotamus limb bones (a–d) and teeth (small open circles), many fragments of bone (short dashes) and a few stone artifacts (colored dots). Trenching (dashed line, color) and hillside excavation over a wide area exposed an ancient soil surface (color) overlying a deposit of silty tuff. Lying on the ancient surface were stone cores (open circles, color) from which sharp-edged flakes had been struck, more than 100 other stone artifacts and more than 60 additional fragments of teeth and bones. The scatter of tools and broken bones suggests the hypothesis that the toolmakers fed on meat from the hippopotamus.

fact association, only a kilometer away from the hippopotamus/artifact site, has allowed us to carry our inquiries further.

The second site had been located by Behrensmeyer in 1969. Erosion was uncovering artifacts, together with pieces of broken-up bone, at another outcrop of the same volcanic ash layer that contained the HAS artifacts and bones. With the assistance of John Barthelme of the University of California at Berkeley and others I began to excavate the site. The work soon revealed a scatter of several hundred stone tools in an area 16 meters in diameter. They rested on an ancient ground surface that had been covered by layers of sand and silt. The concentration of artifacts exactly coincided with a scatter of fragmented bones. Enough of them, teeth in particular, were identifiable to demonstrate that parts of the remains of several animal species were present. John M. Harris of the Louis Leakey Memorial Institute in Nairobi recognized, among other

species, hippopotamus, giraffe, pig, porcupine and such bovids as waterbuck, gazelle and what may be either hartebeest or wildebeest. It was this site that was designated KBS. The site obviously represented the second category of bone-and-artifact associations: tools in association with the remains of many different animal species.

Geological evidence collected by A. K. Behrensmeyer of Yale University and others shows that the KBS deposit had accumulated on the sandy bed of a stream that formed part of a small delta. At the time when the toolmakers used the stream bed, water had largely ceased to flow. Such a site was probably favored as a focus of hominid activity for a number of reasons. First, as every beachgoer knows, sand is comfortable to sit and lie on. Second, by scooping a hole of no great depth in the sand of a stream bed one can usually find water. Third, the growth of trees and bushes in the sun-parched floodplains of East Africa is often densest along watercour-

ses, so that shade and plant foods are available in these locations. It may also be that the protohuman toolmakers who left their discards here took shelter from predators by climbing trees and also spent their nights protected in this way.

Much of this is speculative, of course, but we have positive evidence that the objects at the KBS site did accumulate in the shade. The sandy silts that came to cover the discarded implements and fractured bones were deposited so gently that chips of stone small enough to be blown away by the wind were not disturbed. In the same silts are the impressions of many tree leaves. The species of tree has not yet been formally identified, but Jan Gilette of the Kenya National Herbarium notes that the impressions closely resemble the leaves of African wild fig trees.

Carrying Stones and Meat

As at the hippopotamus/artifact site, we have established the fact that stones

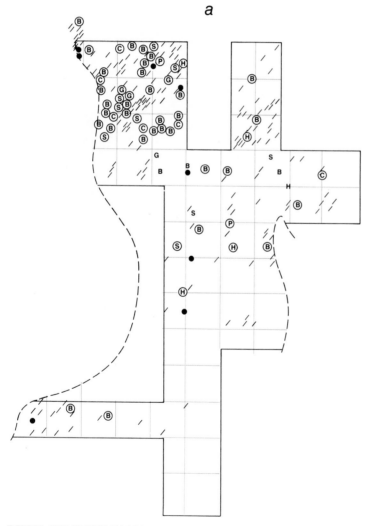

a

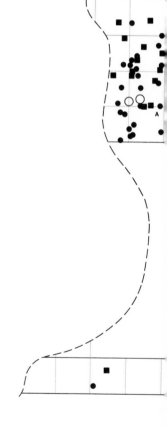

BONES AND STONE TOOLS were also found in abundance at the Kay Behrensmeyer site. As the plot of bone distribution (*a*) shows, the animal remains represent many different species. These are identified by capital letters; if the find was a tooth the letter is circled. Most are small to medium-sized bovids, such as gazelle, waterbuck and hartebeest (*B*). The remains of crocodile (*C*), giraffe (*G*), hippopotamus (*H*), porcupine (*P*) and extinct species of pig (*S*) were also present. Dots and dashes locate unidentified teeth and bone fragments respectively.

larger than the size of a pea do not occur naturally closer to the Kay Behrensmeyer site than a distance of three kilometers. Thus we know that the stones we found at the site must have been carried at least that far. With the help of Frank Fitch and Ron Watkins of the University of London we are searching for the specific sources.

It does not seem likely that all the animals of the different species represented among the KBS bones could have been killed in a short interval of time at this one place. Both considerations encourage the advancement of a tentative hypothesis: Like the stones, the bones were carried in, presumably while there was still meat on them.

If this hypothesis can be accepted, the Kay Behrensmeyer site provides very early evidence for the transport of food as a protohuman attribute. Today the carrying of food strikes us as being commonplace, but as Sherwood Washburn of the University of California at Berkeley observed some years ago such an

action would strike a living ape as being novel and peculiar behavior indeed. In short, if the hypothesis can be accepted, it suggests that by the time the KBS deposit was laid down various fundamental shifts had begun to take place in hominid social and ecological arrangements.

It should be noted that other early sites in this category are known in East Africa, so that the Kay Behrensmeyer site is by no means unique. A number of such sites have been excavated at Olduvai Gorge and reported by Mary Leakey. Of these the best preserved is the "Zinjanthropus" site of Olduvai Bed I, which is about 1.7 million years old. Here too a dense patch of discarded artifacts coincides with a concentration of broken-up bones.

There is an even larger number of Type A sites (where concentrations of artifacts are found but bones are virtually or entirely absent). Some are at Koobi Fora; others are in the Omo Valley, where Harry V. Merrick of Yale Uni-

versity and Jean Chavaillon of the French National Center for Scientific Research (CNRS) have recently uncovered sites of this kind in members E and F of the Shungura Formation. The Omo sites represent the oldest securely dated artifact concentrations so far reported anywhere in the world; the tools were deposited some two million years ago.

One of the Olduvai sites in this category seems to have been a "factory": a quarry where chert, an excellent tool material, was readily available for flaking. The other tool concentrations, with very few associated bones or none at all, may conceivably be interpreted as foci of hominid activity where for one reason or another large quantities of meat were not carried in. Until it is possible to distinguish between sites where bone was never present and sites where the bones have simply vanished because of such factors as decay, however, these deposits will remain difficult to interpret in terms of subsistence ecology.

What, in summary, do these East Af-

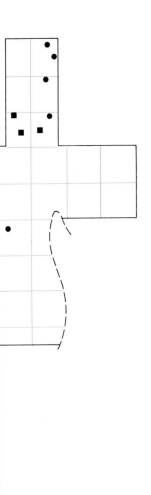

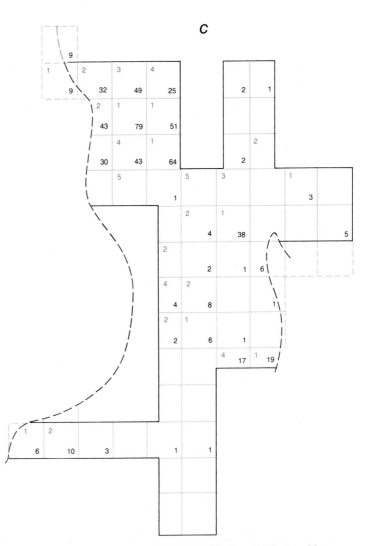

The plot of artifact distribution (*b*) shows that three of four stone cores (*open circles*), most waste stone (*squares*) and flakes and fragments of flakes (*dots*) were found in 12 adjacent squares. Also found here was an unworked stone (*A*) that, like the cores, must have been carried to the site from a distance. Plotting of all tools and bones unearthed at the site was not attempted. Numbers in grid squares (*c*) show how many flakes and bits of waste stone (*color*) and fragments of bone (*black*) were recorded without exact plotting in each square.

rican archaeological studies teach us about the evolution of human behavior? For one thing they provide unambiguous evidence that two million years ago some hominids in this part of Africa were carrying things around, for example stones. The same hominids were also making simple but effective cutting tools of stone and were at times active in the vicinity of large animal carcasses, presumably in order to get meat. The studies strongly suggest that the hominids carried animal bones (and meat) around and concentrated this portable food supply at certain places.

Model Strategies

These archaeological facts and indications allow the construction of a theoretical model that shows how at least some aspects of early hominid social existence may have been organized. Critical to the validity of the model is the inference that the various clusters of re-

mains we have uncovered reflect social and economic nodes in the lives of the toolmakers who left behind these ancient patches of litter. Because of the evidence suggestive of the transport of food to certain focal points, the first question that the model must confront is why early hominid social groups departed from the norm among living subhuman primates, whose social groups feed as they range. To put it another way, what ecological and evolutionary advantages are there in postponing some food consumption and transporting the food?

Several possible answers to this question have been advanced. For example, Adrienne Zihlman and Nancy Tanner of the University of California at Santa Cruz suggest that when the protohumans acquired edible plants out on the open grasslands, away from the shelter of trees, it would have been advantageous for them to seize the plant products quickly and withdraw to places shel-

tered from menacing predators. Others have proposed that when the early hominids foraged, they left their young behind at "nest" or "den" sites (in the manner of birds, wild dogs and hyenas) and returned to these locales at intervals, bringing food with them to help feed and wean the young.

If we look to the recorded data concerning primitive human societies, a third possibility arises. Among extant and recently extinct primitive human societies the transport of food is associated with a division of labor. The society is divided by age and sex into classes that characteristically make different contributions to the total food supply. One significant result of such a division is an increase in the variety of foodstuffs consumed by the group. To generalize on the basis of many different ethnographic reports, the adult females of the society contribute the majority of the "gathered" foods; such foods are mainly plant products but may include shellfish, amphibians and small reptiles, eggs, insects and the like. The adult males usually, although not invariably, contribute most of the "hunted" foodstuffs: the flesh of mammals, fishes, birds and so forth. Characteristically the males and females range in separate groups and each sex eventually brings back to a home base at least the surplus of its foraging.

Could this simple mechanism, a division of the subsistence effort, have initiated food-carrying by early hominids? One cannot dismiss out of hand the models that suggest safety from competitors or the feeding of nesting young as the initiating mechanisms for food-carrying. Nevertheless, neither model seems to me as plausible as one that has division of labor as the primary initiating mechanism. Even if no other argument favored the model, we know for a fact that somewhere along the line in the evolution of human behavior two patterns became established: food-sharing and a division of labor. If we include both patterns in our model of early hominid society, we will at least be parsimonious.

Other arguments can be advanced in favor of an early development of a division of labor. For example, the East African evidence shows that the protohuman toolmakers consumed meat from a far greater range of species and sizes of animals than are eaten by such living primates as the chimpanzee and the baboon. Among recent human hunter-gatherers the existence of a division of labor seems clearly related to the females being encumbered with children, a handicap that bars them from hunting or scavenging, activities that require speed afoot or long-range mobility. For the protohumans too the incorporation of meat in the diet in significant quantities may well have been a key factor in the development not only of a division

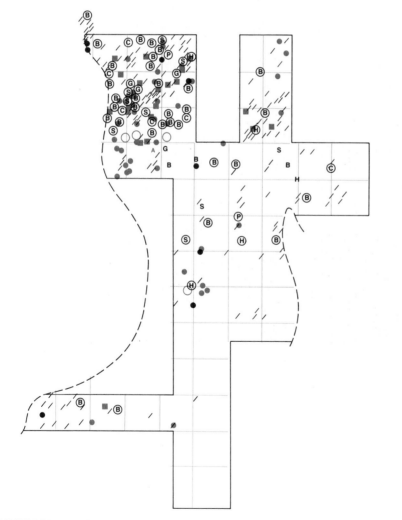

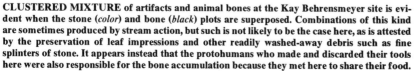

CLUSTERED MIXTURE of artifacts and animal bones at the Kay Behrensmeyer site is evident when the stone (*color*) and bone (*black*) plots are superposed. Combinations of this kind are sometimes produced by stream action, but such is not likely to be the case here, as is attested by the preservation of leaf impressions and other readily washed-away debris such as fine splinters of stone. It appears instead that the protohumans who made and discarded their tools here were also responsible for the bone accumulation because they met here to share their food.

of labor but also of the organization of movements around a home base and the transport and sharing of food.

The model I propose for testing visualizes food-sharing as the behavior central to a novel complex of adaptations that included as critical components hunting and/or scavenging, gathering and carrying. Speaking metaphorically, food-sharing provides the model with a kind of central platform. The adaptive system I visualize, however, could only have functioned through the use of tools and other equipment. For example, without the aid of a carrying device primates such as ourselves or our ancestors could not have transported from the field to the home base a sufficient amount of plant food to be worth sharing. An object as uncomplicated as a bark tray would have served the purpose, but some such item of equipment would have been mandatory. In fact, Richard Borshay Lee of the University of Toronto has suggested that a carrying device was the basic invention that made human evolution possible.

What about stone tools? Our ancestors, like ourselves, could probably break up the body of a small animal, as chimpanzees do, with nothing but their hands and teeth. It is hard to visualize them or us, however, eating the meat of an elephant, a hippopotamus or some other large mammal without the aid of a cutting implement. As the archaeological evidence demonstrates abundantly,

the protohumans of East Africa not only knew how to produce such stone flakes by percussion but also found them so useful that they carried the raw materials needed to make the implements with them from place to place. Thus whereas the existence of a carrying device required by the model remains hypothetical as far as archaeological evidence is concerned, the fact that tools were used and carried about is amply attested to.

In this connection it should be stressed that the archaeological evidence is also silent with regard to protohuman consumption of plant foods. Both the morphology and the patterns of wear observable on hominid teeth suggest such a plant component in the diet, and so does the weight of comparative data on subsistence patterns among living nonhuman primates and among nonfarming human societies. Nevertheless, if positive evidence is to be found, we shall have to sharpen our ingenuity, perhaps by turning to organic geochemical analyses. It is clear that as long as we do not correct for the imbalance created by the durability of bone as compared with that of plant residues, studies of human evolution will tend to have a male bias!

As far as the model is concerned the key question is not whether collectable foods—fruits, nuts, tubers, greens and even insects—were eaten. It is whether these protohumans carried such foods about. Lacking any evidence for the

consumption of plant foods, I shall fall back on the argument that the system I visualize would have worked best if the mobile hunter-scavenger contribution of meat to the social group was balanced by the gatherer-carrier collection of high-grade plant foods. What is certain is that at some time during the past several million years just such a division of labor came to be a standard kind of behavior among the ancestors of modern man.

A final cautionary word about the model: The reader may have noted that I have been careful about the use of the words "hunter" and "hunting." This is because we cannot judge how much of the meat taken by the protohumans of East Africa came from opportunistic scavenging and how much was obtained by hunting. It is reasonable to assume that the carcasses of animals killed by carnivores and those of animals that had otherwise died or been disabled would always have provided active scavengers a certain amount of meat. For the present it seems less reasonable to assume that protohumans, armed primitively if at all, would be particularly effective hunters. Attempts are now under way, notably by Elizabeth Vrba of South Africa, to distinguish between assemblages of bones attributable to scavenging and assemblages attributable to hunting, but no findings from East Africa are yet available. For the present I am inclined to accept the verdict of J. Desmond

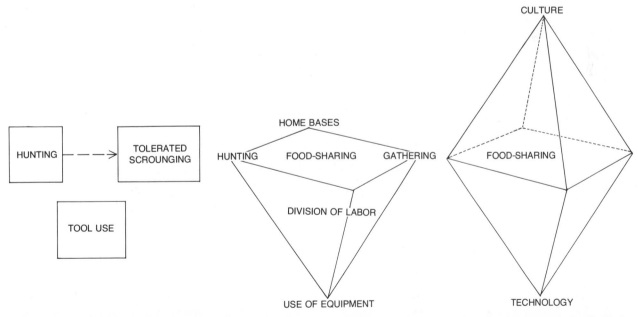

BEHAVIOR PATTERNS that differ in degree of organization are contrasted in these diagrams. Living great apes, exemplified here by the chimpanzee, exhibit behavior patterns that became important in human evolution but the patterns (*left*) exist largely as isolated elements. Hunting occurs on a small scale but leads only to "tolerated scrounging" rather than active food-sharing; similarly, tools are used but tool use is not integrated with hunting or scrounging. The author's model (*center*) integrates these three behavior patterns and others into a coherent structure. Food-sharing is seen as a central structural element, incorporating the provision of both animal and plant foods, the organization of a home base and a division of labor. Supporting the integrated structure is a necessary infrastructure of tool and equipment manufacture; for example, without devices for carrying foodstuffs there could not be a division of labor and organized food-sharing. In modern human societies (*right*) the food-sharing structure has undergone socioeconomic elaboration. Its infrastructure now incorporates all of technology, and a matching superstructure has arisen to incorporate other elements of what is collectively called culture.

Clark of the University of California at Berkeley and Lewis R. Binford of the University of New Mexico. In their view the earliest meat-eaters might have obtained the flesh of animals weighing up to 30 kilograms by deliberate hunting, but the flesh of larger animals was probably available only through scavenging.

Tools as Testimony

Of course, the adaptive model I have advanced here reflects only a working hypothesis and not established fact. Nevertheless, there is sufficient evidence in its favor to justify looking further at its possible implications for the course of human evolution. For example, the model clearly implies that early toolmaking hominids displayed certain patterns of behavior that, among the patterns of behavior of all primates, uniquely characterize our own species and set it apart from its closest living relatives, the great apes. Does this mean that the toolmaking hominids of 1.5 to two million years ago were in fact "human"?

I would surmise that it does not, and I have been at pains to characterize these East African pioneers as protohumans. In summarizing the contrasts between living men and living apes I put high on the list language and the cultural phenomena that are dependent on it. We have no direct means of learning whether or not any of these early hominids had language. It is my suspicion, however, that the principal evolutionary change in the hominid line leading to full humanity over the past two million years has been the great expansion of language and communication abilities, together with the cognitive and cultural capabilities integrally related to language. What is the evidence in support of this surmise?

One humble indicator of expanding mental capacities is the series of changes that appears in the most durable material record available to us: the stone tools. The earlier tools from the period under consideration here seem to me to show a simple and opportunistic range of forms that reflect no more than an uncomplicated empirical grasp of one skill: how to fracture stone by percussion in such a way as to obtain fragments with sharp edges. At that stage of toolmaking the maker imposed a minimum of culturally dictated forms on his artifacts. Stone tools as simple as these perform perfectly well the basic functions that support progress in the direction of becoming human, for example the shaping of a digging stick, a spear and a bark tray, or the butchering of an animal carcass.

The fact is that exactly such simple stone tools have been made and used ever since their first invention, right down to the present day. Archaeology also shows, however, that over the past several hundred thousand years some assemblages of stone tools began to reflect a greater cultural complexity on the part of their makers. The complexity is first shown in the imposition of more arbitrary tool forms; these changes were followed by increases in the number of such forms. There is a marked contrast between the pure opportunism apparent in the shapes of the earliest stone tools and the orderly array of forms that appear later in the Old Stone Age when each form is represented by numerous standardized examples in each assemblage of tools. The contrast strongly suggests that the first toolmakers lacked the highly developed mental and cultural abilities of more recent humans.

The evidence of the hominid fossils and the evidence of the artifacts together suggest that these early artisans were nonhuman hominids. I imagine that if we had a time machine and could visit a place such as the Kay Behrensmeyer site at the time of its original occupation, we would find hominids that were living in social groups much like those of other higher primates. The differences would be apparent only after prolonged observation. Perhaps at the start of each day we would observe a group splitting up as some of its members went off in one direction and some in another. All these subgroups would very probably feed intermittently as they moved about and encountered ubiquitous low-grade plant foods such as berries, but we might well observe that some of the higher-grade materials—large tubers or the haunch of a scavenged carcass—were being reserved for group consumption when the foraging parties reconvened at their starting point.

To the observer in the time machine behavior of this kind, taken in context with the early hominids' practice of making tools and equipment, would seem familiarly "human." If, as I suppose, the hominids under observation communicated only as chimpanzees do or perhaps by means of very rudimentary protolinguistic signals, then the observer might feel he was witnessing the activities of some kind of fascinating bipedal ape. When one is relying on archaeology to reconstruct protohuman life, one must strongly resist the temptation to project too much of ourselves into the past. As Jane B. Lancaster of the University of Oklahoma has pointed out, the hominid life systems of two million years ago have no living counterparts.

Social Advances

My model of early hominid adaptation can do more than indicate that the first toolmakers were culturally protohuman. It can also help to explain the dynamics of certain significant advances in the long course of mankind's development. For example, one can imagine that a hominid social organization involving some division of labor and a degree of food-sharing might well have been able to function even if it had communicative abilities little more advanced than those of living chimpanzees. In such a simple subsistence system, however, any group with members that were able not only to exchange food but also to exchange information would have gained a critical selective advantage over all the rest. Such a group's gatherers could report on scavenging or hunting opportunities they had observed, and its hunters could tell the gatherers about any plant foods they had encountered.

By the same token the fine adjustment of social relations, always a matter of importance among primates, becomes doubly important in a social system that involves food exchange. Language serves in modern human societies not only for the exchange of information but also as an instrument for social adjustment and even for the exchange of misinformation.

Food-sharing and the kinds of behavior associated with it probably played an important part in the development of systems of reciprocal social obligations that characterize all human societies we know about. Anthropological research shows that each human being in a group is ordinarily linked to many other members of the group by ties that are both social and economic. The French anthropologist Marcel Mauss, in a classic essay, "The Gift," published in 1925, showed that social ties are usually reciprocal in the sense that whereas benefits from a relationship may initially pass in only one direction, there is an expectation of a future return of help in time of need. The formation and management of such ties calls for an ability to calculate complex chains of contingencies that reach far into the future. After food-sharing had become a part of protohuman behavior the need for such an ability to plan and calculate must have provided an important part of the biological basis for the evolution of the human intellect.

The model may also help explain the development of human marriage arrangements. It assumes that in early protohuman populations the males and females divided subsistence labor between them so that each sex was preferentially tapping a different kind of food resource and then sharing within a social group some of what had been obtained. In such circumstances a mating system that involved at least one male in "family" food procurement on behalf of each child-rearing female in the group would have a clear selective advantage over, for example, the chimpanzees' pattern of opportunistic relations between the sexes.

I have emphasized food-sharing as a principle that is central to an understanding of human evolution over the past two million years or so. I have also set forth archaeological evidence that food-sharing was an established kind of behavior among early protohumans. The notion is far from novel; it is implicit in many philosophical speculations and in many writings on paleoanthropology. What is novel is that I have undertaken to make the hypothesis explicit so that it can be tested and revised.

Accounting for Evolution

Thus the food-sharing hypothesis now joins other hypotheses that have been put forward to account for the course of human evolution. Each of these hypotheses tends to maintain that one or another innovation in protohuman behavior was the critical driving force of change. For example, the argument has been advanced that tools were the "prime movers." Here the underlying implication is that in each successive generation the more capable individuals made better tools and thereby gained advantages that favored the transmission of their genes through natural selection; it is supposed that these greater capabilities would later be applied in aspects of life other than technology. Another hypothesis regards hunting as being the driving force. Here the argument is that hunting requires intelligence, cunning, skilled neuromuscular coordination and, in the case of group hunting, cooperation. Among other suggested prime movers are such practices as carrying and gathering.

If we compare the food-sharing explanation with these alternative explanations we see that in fact food-sharing incorporates many aspects of each of the others. It will also be seen that in the food-sharing model the isolated elements are treated as being integral parts of a complex, flexible system. The model itself is probably an oversimplified version of what actually happened, but it seems sufficiently realistic to be worthy of testing through further archaeological and paleontological research.

Lastly, the food-sharing model can be seen to have interconnections with the physical implications of fossil hominid anatomy. For example, a prerequisite of food-sharing is the ability to carry things. This ability in turn is greatly facilitated by a habitual two-legged posture. As Gordon W. Hewes of the University of Colorado has pointed out, an important part of the initial evolutionary divergence of hominids from their primate relatives may have been the propensity and the ability to carry things about. To me it seems equally plausible that the physical selection pressures that promoted an increase in the size of the protohuman brain, thereby surely enhancing the hominid capacity for communication, are a consequence of the shift from individual foraging to food-sharing some two million years ago.

BIBLIOGRAPHIES

1. Tools and Human Evolution

CEREBRAL CORTEX OF MAN. Wilder Penfield and Theodore Rasmussen. The Macmillan Company, 1950.

THE EVOLUTION OF MAN, edited by Sol Tax. University of Chicago Press, 1960.

THE EVOLUTION OF MAN'S CAPACITY FOR CULTURE. Arranged by J. N. Spuhler. Wayne State University Press, 1959.

HUMAN ECOLOGY DURING THE PLEISTOCENE AND LATER TIMES IN AFRICA SOUTH OF THE SAHARA. J. Desmond Clark in *Current Anthropology*, Vol. I, pages 307–324; 1960.

2. The Early Relatives of Man

A CRITICAL REAPPRAISAL OF TERTIATY PRIMATES. Elwyn L. Simons in *Evolutionary and Genetic Biology of Primates: Vol. I.* Academic Press, 1963.

THE MIOCENE HOMINOIDEA OF EAST AFRICA. W. E. Le Gros Clark and L. S. B. Leakey in *Fossil Mammals of Africa: No. I, British Museum (Natural History)*, pages 1–115; 1951.

A REVIEW OF THE MIDDLE AND UPPER EOCENE PRIMATES OF NORTH AMERICA. C. Lewis Gazin in *Smithsonian Miscellaneous Collections*, Vol. 136, No. 1, pages 1–112; July, 1958.

SOME FALLACIES IN THE STUDY OF HOMINID PHYLOGENY. Elwyn L. Simons in *Science*, Vol. 141, No. 3584, pages 879–889; September, 1963.

3. Ramapithecus

NEWLY RECOGNIZED MANDIBLE OF *RAMAPITHECUS*. David Pilbeam in *Nature*, Vol. 222, No. 5198, pages 1093–1094; June 14, 1969.

RAMAPITHECUS in *Primate Evolution: An Introduction to Man's Place in Nature.* Elwyn L. Simons. The Macmillan Company, 1972.

THE PRIMATE AND OTHER FAUNA FROM FORT TERNAN, KENYA. Peter Andrews and Alan Walker in *Human Origins: Louis Leakey and the East African Evidence*, edited by Glynn Ll. Isaac and Elizabeth R. McCown. W. A. Benjamin, Inc., 1976.

4. The Evolution of the Hand

THE ANTECEDENTS OF MAN: AN INTRODUCTION TO THE EVOLUTION OF THE PRIMATES. W. E. Le Gros Clark. Edinburgh University Press, 1959.

MAN THE TOOL-MAKER. Kenneth P. Oakley. British Museum of Natural History, 1950.

THE PREHENSILE MOVEMENTS OF THE HUMAN HAND. J. R. Napier in *The Journal of one and Joint Surgery*, Vol. 38-B, No. 4, pages 902–213; November, 1956.

PREHENSILITY AND OPPOSABILITY IN THE HANDS OF PRIMATES. J. R. Napier in *Symposia of the Zoological Society of London*, No. 5, pages 115–132; August, 1961.

THE TROPICAL RAIN FOREST. P. W. Richards. Cambridge University Press, 1957.

5. The Antiquity of Human Walking

THE APE-MEN. Robert Broom in *Scientific American;* November, 1949.

THE FOOT AND THE SHOE. J. R. Napier in *Physiotherapy*, Vol. 43, No. 3, pages 65–74; March, 1957.

A HOMINID TOE BONE FROM BED 1, OLDUVAI GORGE, TANZANIA. M. H. Day and J. R. Napier in *Nature*, Vol. 211, No. 5052, pages 929–930, August 27, 1966.

6. The Hominids of East Turkana

EARLIEST MAN AND ENVIRONMENTS IN THE LAKE RUDOLF BASIN: STRATIGRAPHY, PALEOECOLOGY AND EVOLUTION. Edited by Yves Coppens. F. C. Howell,

Glynn Ll. Isaac and Richard E. F. Leakey. University of Chicago Press, 1976.

HUMAN ORIGINS: LOUIS LEAKEY AND THE EAST AFRICAN EVIDENCE. Edited by Glynn Ll. Isaac and Elizabeth R. McCown. W. A. Benjamin, Inc., 1976.

KOOBI FORA RESEARCH PROJECT, Vol. 1: THE FOSSIL HOMINIDS AND AN INTRODUCTION TO THEIR CONTEXT 1968–1974. Edited by Mary Leakey and Richard E. F. Leakey. Oxford University Press, 1978.

7. The Casts of Fossil Hominid Brains

THE EVOLUTION OF THE PRIMATE BRAIN: SOME ASPECTS OF QUANTITATIVE RELATIONSHIPS. R. L. Holloway in *Brain Research*, Vol. 7, pages 121–172; 1968.

AUSTRALOPHITHECENE ENDOCASTS, BRAIN EVOLUTION IN THE HOMINOIDEA, AND A MODEL OF HOMINID EVOLUTION. R. L. Holloway in *The Functional and Evolutionary Biology of Primates*, edited by R. Tuttle. Aldine Press, 1972.

EVOLUTION OF PRIMATE BRAINS: A COMPARATIVE ANATOMICAL INVESTIGATION. H. Stephan in *The Functional and Evolutionary Biology of Primates*, edited by R. Tuttle. Aldine Press, 1972.

8. Homo Erectus

MANKIND IN THE MAKING. William W. Howells. Doubleday & Company, Inc., revised edition, 1967.

THE NOMENCLATURE OF THE HOMINIDAE. Bernard G. Campbell, Occasional Paper No. 22, Royal Anthropological Institute of Great Britain and Ireland, 1965.

THE TAXONOMIC EVOLUTION OF FOSSIL HOMINIDS. Ernst Mayr in *Classification and Human Evolution*, edited by Sherwood L. Washburn. Viking Fund Publications in Anthropology, No. 37, 1963.

9. Stone Tools and Human Behavior

LE PALÉOLITHIQUE INFÉRIEUR ET MOYEN DE JABRUD (SYRIE) ET LA QUESTION DUE PRÉ-AURIGNACIEN. F. Bordes in *L'Anthropologie*, Vol. 59, Nos. 5–6, pages 486–507; 1955.

THE HUMAN USE OF THE EARTH. Philip L. Wagner. The Free Press of Glencoe, Illinois, 1960.

ME'ARAT SHOVAKH (MUGHARET ES-SHUBBABIQ). Sally R. Binford in *Israel Exploration Journal*, Vol. 16, No. 1, pages 18–32, No. 2, pages 96–103; 1966.

A PRELIMINARY ANALYSIS OF FUNCTIONAL VARIABILITY IN THE MOUSTERIAN OF LEVALLOIS FACIES. Lewis R. Binford and Sally R. Binford in *American Anthropologist*, Vol. 68, No. 2, Part 2, pages 238–295; April, 1966.

MODERN FACTOR ANALYSIS. Harry H. Harman. The University of Chicago Press, 1967.

10. The Functions of Paleolithic Flint Tools

PREHISTORIC TECHNOLOGY: AN EXPERIMENTAL STUDY OF THE OLDEST TOOLS AND ARTEFACTS FROM TRACES OF MANUFACTURE AND WEAR. S. A. Semenov, translated and with an introduction by M. W. Thompson. Barnes & Noble, Inc., 1964.

MICROWEAR ANALYSIS OF EXPERIMENTAL FLINT TOOLS: A TEST CASE. Lawrence H. Keeley and Mark H. Newcomer in *Journal of Archaeological Science*, Vol. 4, No. 1, pages 29–62; March, 1977.

11. The Food-Sharing Behavior of Protohuman Hominids

PRIMATE BEHAVIOR AND THE EMERGENCE OF HUMAN CULTURE. Jane B. Lancaster. Holt, Rinehart & Winston, 1975.

EARLIEST MAN AND ENVIRONMENTS IN THE LAKE RUDOLF BASIN: STRATIGRAPHY, PALEOECOLOGY AND EVOLUTION. Edited by Yves Coppens, F. C. Howell, Glynn Ll. Isaac and Richard E. F. Leakey. University of Chicago Press, 1976.

HUMANKIND EMERGING. Edited by Bernard G. Campbell. Little, Brown & Company, 1976.

HUMAN ORIGINS: LOUIS LEAKEY AND THE EAST AFRICAN EVIDENCE. Edited by Glynn Ll. Isaac and Elizabeth R. McCown. W. A. Benjamin, Inc., 1976.

INDEX